Multi-factor Models and Signal Processing Techniques

Multi-factor Models and Signal Processing Techniques

Application to Quantitative Finance

Serge Darolles
Patrick Duvaut
Emmanuelle Jay

WILEY

First published 2013 in Great Britain and the United States by ISTE Ltd and John Wiley & Sons, Inc.

ISTE Ltd
27-37 St George's Road
London SW19 4EU
UK

www.iste.co.uk

John Wiley & Sons, Inc.
111 River Street
Hoboken, NJ 07030
USA

www.wiley.com

Library of Congress Control Number: 2013941318

British Library Cataloguing-in-Publication Data
A CIP record for this book is available from the British Library
ISBN: 978-1-84821-419-4

Printed and bound in Great Britain by CPI Group (UK) Ltd., Croydon, Surrey CR0 4YY

Table of Contents

Foreword

With the irruption of electronic markets and the widespread use of quantitative models for trading, order execution and risk management, financial markets have evolved from the low-tech environment where they started from a few decades ago to the one where market participants routinely employ advanced data processing techniques to analyze large data flows in real time and feed the results into trading algorithms and risk management systems. Operational constraints arising from the sheer volume of data to be analyzed have created the need for intelligent data processing and modeling methods and parsimonious models whose estimation and simulation is feasible in the high-dimensional settings represented by the large number of risk factors involved in risk decisions.

In this context, advanced tools and methods from signal processing – the branch of science and engineering that deals with the modeling and analysis of complex streams of data – are increasingly used to analyze financial data and estimate the statistical models used by portfolio managers and risk managers.

In contrast to the traditional toolkit of portfolio theory, which is mostly based on *static* analysis of portfolio loss distributions, the use of models based on stochastic process allows risk managers a dynamic, and more realistic, view of portfolio risk which is dynamically updated as new information is revealed. However, the use of such dynamic models comes at a price: we need more advanced tools to estimate, simulate and analyze portfolio risk in these models. Tools that have long been used in engineering, such as the Kalman filter, but may be unfamiliar to some risk management professionals, turn out to be relevant when moving beyond the static setting.

Simultaneously, the use of high-dimensional models entail new estimation problems: in many asset allocation problems, one has a few years (so, a few hundreds of daily observations) to estimate the parameters describing the evolution of hundreds, if not thousands, of risk factors. In this context, traditional estimation methods, such as principal component analysis and basic covariance analysis, fail to be consistent and yield unexpected (and biased!) results ... Tools such as random matrix theory, initially developed in nuclear physics to model the energy levels of complex nuclei, have been recently used to explore such estimation biases and propose methods for overcoming them.

Leveraging on their double profile of quantitative risk professionals with a strong research and technical background, Serge Darolles, Patrick Duvaut and Emmanuelle Jay have done an excellent job in presenting a nice selection of advanced signal processing tools and statistical models that address some of these points in a self-contained manner, at a level accessible to students and professionals involved with quantitative risk management.

By drawing on their professional experience in quantitative modeling to achieve the right level of technical sophistication, Serge, Patrick and Emmanuelle have succeeded in covering a variety of relevant concepts without sacrificing the technical details necessary for a serious implementation: multi-factor models, advanced estimation methods based on random matrix theory, filtering algorithms and regularization methods for identification of latent risk factors. The use of these tools in risk management and asset allocation is nicely illustrated through examples using financial data.

I trust that the hard work done by the authors will benefit a wide range of readers across academic and professional circles and act as an incentive for readers to delve into the vast research literature on statistical modeling in finance. *Bonne lecture!*

Rama CONT
London
June 2013

Introduction

I.1. Digital society and new paradigms in fund management

The massive digitization of all our activities from entertainment to all public and economical domains marks the era of the digital society. This latter generates an overwhelming amount of data, measured in zetta (10^{21}) bytes.

This phenomenon called *Big Data* impacts the finance sphere as well. The so-called *five Vs* need to be handled by fund management companies: volume, velocity, variety, visualization and value. Moreover, the 2008 mortgage crisis and its macroeconomic effects had two major consequences that make the situation even more difficult for fund managers. First, these large amounts of data are highly *hectic* (non-stationary), even more unpredictable and noisy. Second, the fund management paradigm has been shifted from an "abstruse gain race" at any risk to being able to control the risks (draw down) while securing a minimum gain, with a maximum transparency.

One way to address the above mentioned challenging issues is to provide fund managers with appropriate and "user friendly" QUANT(ITATIVE) tools that are market driven. This is where Factor Models, Statistical Signal Processing (SSP) and Multi-factor Models and Signal Processing Techniques come into play.

I.2. Statistical signal processing and factor models in finance

The linear factor model is nowadays a benchmark in portfolio management theory and in the understanding of asset returns, even in the hedge fund industry where returns may have high nonlinearities. The idea is to relate a large number and variety of returns to a few relevant *factors*, up to Gaussian errors that cannot be described solely by the factors. The factors thus drive and parsimoniously explain the *spatial correlation between* assets, while the errors are *idiosyncratic to each* asset.

Factors need to be chosen and/or identified first, then the parameters of the linear model that connect them to the assets, such as the *alphas* (gains) and the *betas* (assets exposures) have to be estimated and tracked. Based on a linear regression fed by market data, linear factor models are thus suitable for all the requirements listed above of friendliness to the users, transparency and as being market driven.

Statistical signal processing (SSP) methods allow us to access both the factors and the model parameters, in a very noisy, highly unpredictable and hectic environment. SSP methods have proved their ergonomy, efficiency and practicality in highly hostile and changing conditions in the past few decades when applied to medical diagnostics, non-destructive testing of nuclear plants, aircrafts, cars and in boosting both the performance and the robustness of mobile phones, with some top achievements in the 4G Long Term Evolution (LTE) standards.

The aim of this book is to share the same when dealing with fund management and more specifically when facing the identification of linear factor models. While application of SSP to finance has been ongoing for a few years, it is still a big challenge due to the difficulty of relying on joint expertise in fund management, statistical signal processing and being at the crossroad of Academia and Business, to make the recent results of research usable for fund managers.

I.3. Multiexpertise team of authors

The three authors' complementary expertise and experience meet the challenging above mentioned conditions. First, the three of them have both academia and business experience. Second, two of them are experienced fund managers, while the third was an executive of a US company involved in asset capitalization. Moreover, the three authors have been working together many years, as partners of the same Company QAMLab©, involved in quantitative solutions for fund management. It is worth mentioning that the authors' expertise range from mathematical finance to SSP.

This book is thus the outcome of such a unique pluridisciplinary and market-oriented expertise.

I.4. Book positioning and targeted readers

The book targets three worldwide audiences. First, graduate students of Business, Finance, Management Masters, including, of course, MBAs who would like to understand how SSP methods can leverage linear factor models in the complicated and hectic digital asset management activities of the emerging digital society. Second, SSP grad students who would like to ramp up in quantitative finance following familiar tracks of least squares, Kalman filtering (KF) and more advanced tools based on most recent regularized Kalman approaches. Third, qualitative fund managers who are eager to add to their daily management softwares and practices, several standalone and/or end-to-end SSP methods in order to enhance and boost their own management style, in a very practical, transparent and market-oriented manner.

The book material is unique since it presents QUANT tools as practical answers to market driven issues faced by fund managers. First in the associated chapters, the SSP material is introduced from the users' stand-point, answering the questions, why do we need these tools, how do they work, how do we use them, how can they be tuned to the real data? The derivations of the results are provided in appendices and can be skipped for user-only readers. On the other hand, SSP skillful audience can access quickly this advanced material. Second, numerous examples and real market data are used through the whole book to better understand the SSP methods and their pluses when dealing with linear factor models. Finally, highlights emphasize the most relevant contents at the end of each chapter, to allow quick access.

It is important to note that the book contents were already taught by all the authors, both in finance and Information Science Masters. Feedback from the students has thus been a valuable input to improve the clarity and the depth of the writing.

I.5. Book organization

The book is organized into four chapters.

Chapter 1 introduces the factor model foundations of the whole book, the equations, the notations, the main concepts, the origin, the applications, etc. While most of it can be skipped by readers familiar with fund management, it is nevertheless valuable to check it to master the model that will be used in the remainder of the book.

Chapter 2 presents qualitative and quantitative ways to choose the factors. While qualitative approaches leverage pure markets know-how, quantitative methods are based on the so-called *eigenfactors*, derived from the eigen-decomposition of the returns covariance matrix. Statistical methods to select the "optimum" number of factors are also detailed. Even SSP readers can enrich their knowledge by grasping the connection between eigen-methods and the determination of the appropriate number of factors.

Least squares and Kalman algorithms that aim to estimate the alphas and the betas of the factor model are the main focus of Chapter 3. A geometrical presentation of both methods enables us to understand their objectives, usage, performance, and limitations in a practical way, especially to the attention of readers from finance and business backgrounds. The whole derivation of KF is provided in Chapter 3 appendix, still along a geometrical fashion. These two methods are basic tools in SSP.

Chapter 4 contents are the most advanced from SSP standpoint and share results at the edge (at the time the book is written) of relevant research about KF regularization to handle some impulsive and non-Gaussian markets behaviors. Spectacular improvement with respect to traditional KF is obtained and illustrated on market data.

Notations and Acronyms

a, **b**, etc.	Column vectors
A, **B**, etc.	Matrices
$\mathbf{A}'$	Transpose of matrix **A**
$\mathbf{A}^H$	Transpose conjugate of matrix **A**
$\mathbf{A}^{-1}$	Inverse of matrix **A**
$\text{tr}(\mathbf{A})$	Trace of matrix **A**
$\nabla_{\mathbf{x}}(.)$	Gradient vector-operator
$\mathbf{0}_N$	$N \times N$ null matrix
$\mathbf{I}_N$	$N \times N$ identity matrix
$\mathbf{1}_N$	N-dimensional vector of 1s
a, b, etc.	Scalars
t	Time
T	Total number of time observations
$t = 1, \cdots, T$	One-step vector of time going from 1 to T by step of 1
N	Number of assets
$s_{i,t}$	Price of asset i at time t
r_t	Return of an asset at time t
r_f	Return of the risk-free asset
$r_{i,t}$	Return of asset i at time t
$r_{m,t}$	Return of the market m at time t
$\mathbf{r}_t$	N-dimensional vector of returns at time t
$\mathbf{r}_k$	T-dimensional vector of returns for asset k
$\mathbf{R} = \{\mathbf{r}_k\}_{k=1}^N$	Matrix of returns for the N observed assets

$\mathbf{F} = \{\mathbf{f}_k\}_{k=1}^{K}$	Matrix of the K factor returns included in factor models
$\mathbf{f}_k$	kth column of $\mathbf{F}$
$\mathbf{f}_t$	tth row of $\mathbf{F}$ and K-dimensional vector of the factor values at t
$\mathbf{G} = \left[\mathbf{1}\ \mathbf{F}\right]$	Matrix $\mathbf{F}$ augmented by a vector of 1s having the same number of rows
$\mathbf{g}_k$	kth column of $\mathbf{G}$
$\mathbf{g}_t$	tth row of $\mathbf{G}$ and $K+1$-dimensional vector of values at t
α	"Alpha" or the specific performance of an asset over the market performance
β	"Beta" or the exposure of an asset to a specific factor
$\boldsymbol{\beta}$ or b	"Betas" or the vector of asset exposures to the K selected factors
$\boldsymbol{\theta} = \left[\alpha\ \boldsymbol{\beta}^T\right]^T$	$K+1$-dimensional vector of the unknown parameters of a multi-factor model
$\lVert\cdot\rVert_q$	l^q-norm
$\mathbb{1}(.)$	Indicator function
$\hat{a}$	Estimated value of a
i.i.d.	Independent and identically distributed
$\sim$	Distributed according to
$\perp\!\!\!\perp$	Statistically independent
$\mathbb{E}(.)$	Statistical expectation
$V(.)$ or $Var(.)$	Variance
$Cov(.)$	Covariance
$\mathcal{N}(\mu,\sigma^2)$	Gaussian distribution with mean μ and variance σ^2
$\mathcal{N}(\boldsymbol{\mu},\boldsymbol{\Sigma})$	Multivariate Gaussian distribution with a mean vector $\boldsymbol{\mu}$ and covariance $\boldsymbol{\Sigma}$
AIC	Akaike information criterion
APT	Arbitrage pricing theory
BIC	Bayes information criterion
CAPM	Capital asset pricing model

FP	Fixedpoint
KF	Kalman filter
LSE	Least squares estimate
MDL	Minimum description length
ML	Maximum likelihood
MMSE	Minimum mean square error
MSE	Mean square error
MUSIC	Multiple signal characterization
NAV	Net asset value
OLS	Ordinary least squares
PCA	Principal component analysis
QAMLAB©	Quantitative Asset Management Laboratory
rgKF	Regularized Kalman filter
RKF	Robust Kalman filter
RMSE	Root mean square error
RMT	Random matrix theory
SCM	Sample covariance matrix
SSP	Statistical signal processing
SW-OLS	Sliding window ordinary least squares

Chapter 1

Factor Models and General Definition

1.1. Introduction

Today, the linear factor model is a benchmark in portfolio management theory [MAR 52, LIN 65], and arbitrage pricing theory (APT) [ROS 76, ROL 80]. In practice, factor models are also used widely to understand the cross-section dispersion of asset returns, whatever the asset class is. Even in the hedge fund industry, where returns feature high nonlinearity, this approach is largely implemented, in general with some improvement such as nonlinear factors or time-varying parameters. This chapter introduces not only the common version of linear factor models but also discusses its limits and the developments described in the following chapters of this book.

In section 1.2, we introduce the different notations and discuss the model and its structure. We list in section 1.3 the reasons why factor models are generally used in finance, and discuss the limits of this approach. Section 1.4 describes the different steps in the building of factor models, i.e. factor selection and parameter estimation. This section is a direct introduction to Chapter 2 for the factor selection step, and Chapters 3 and 4 for the parameter estimation step. Finally, section 1.5 concludes the chapter by giving a historical perspective on the use of factor models in finance.

1.2. What are factor models?

1.2.1. *Notations*

We first consider a set of N risky assets, indexed $i = 1, ..., N$. We denote by $s_{i,t}$ the price of the asset i at time t, and $r_{i,t}$ the corresponding return for the period $(t-1, t)$. This return is defined by:

$$r_{i,t} = \frac{s_{i,t}}{s_{i,t-1}} - 1. \qquad [1.1]$$

If prices are observed on a daily (respectively, weekly, monthly, etc.) basis, then $r_{i,t}$ represents daily (respectively, weekly, monthly, etc.) returns. We then denote by μ_i the expected return of asset i, and σ_i^2 its variance:

$$\mu_i = \mathbb{E}(r_{i,t}) \text{ and } \sigma_i^2 = V(r_{i,t}). \qquad [1.2]$$

The variance σ_i^2 (or the volatility σ_i) is often used to measure the risk relative to the asset i. The greater the volatility, the greater is the risk. The covariance $\sigma_{i,j} = Cov(r_{i,t}, r_{j,t})$ between assets i and j will be useful to compute the risk of a portfolio of several assets.

As opposed to risky assets, we also consider a risk-free asset that gives a return r_f, called the risk-free rate (supposed to be constant). In practice, short-term government securities such as US Treasury Bills are used as a proxy for the risk-free asset; this results in a non-constant r_f.

The expected excess return (or risk premium) of asset i is the return we can expect from asset i in excess of r_f, that is

$$\mathbb{E}(r_{i,t}) - r_f, \qquad [1.3]$$

and is the premium relative to the risk taken when investing in asset i.

According to Markowitz's [MAR 52] theory of *mean-variance efficiency*, investors (should) therefore require a higher expected return for holding a more risky asset and want to earn the highest possible return for a level of risk that they are willing to take. A portfolio is said to be mean-variance efficient if we cannot create another portfolio with a greater expected return and the same variance of returns (or another portfolio with a lower variance and the same expected return).

Figure 1.1 illustrates the time series evolution of three risky assets in comparison to an asset that is a proxy for the risk-free asset. Their mean-variance trade-offs are given in Table 1.1.

Name	$\mu(\%)$	$\sigma(\%)$	$(\mu - r_f)/\sigma$
T-Bill 1Y (r_f)	2.93	1.19	-
HFRI ED Index	8.91	7.1	0.84
S&P500	4.28	14.69	0.09
APPLE	58.3	41.05	1.35

Table 1.1. *Annualized return (μ), annualized volatility (σ) and return-to-risk ratio (Sharpe Ratio) calculated between Feburary 2003 and November 2009 for the four financial instruments shown in Figure 1.1 and ordered by increasing level of risk*

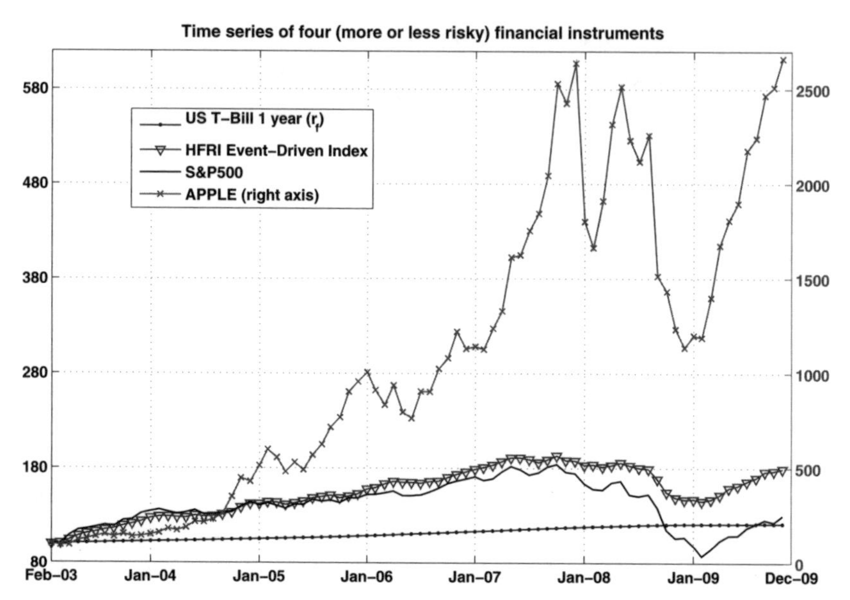

Figure 1.1. *Monthly evolution of four time series (in US dollars and based at $100 at the end of January 2003) representing: (1) an investment in the US T-Bill 1 year, a 1 year maturity US Treasury Bill, considered as a proxy for the risk-free asset, (2) the NAV of HFRI ED index, an index of event-driven (ED) hedge funds, (3) the price of the S&P500, a US market index and (4) the price of the equity APPLE (on the right axis)*

The net asset value (NAV) gives the evolution over time of $100 invested in each asset. We can easily compute the NAV from the arithmetic returns with:

$$NAV_{i,t} = NAV_{i,t-1}\left(1 + r_{i,t}\right), \forall i = 1, \cdots, N \text{ and } \forall t = 1, \cdots, T$$

with $NAV_{i,0} = 100$ (end of January 2003).

1.2.2. *Factor representation*

A factor model is a multivariate regression linking the returns of a set of risky assets to several factors. We focus our attention on linear factor models where the relationship between factors and returns is linear.

Factors represent fundamental data, statistical factors or specific portfolios. Fundamental data are specified using economic theory and the knowledge of financial markets and include macroeconomic variables such as the inflation rate, the unemployment rate and the gross national product. Typically, macroeconomic variables are correlated.

To select uncorrelated factors, empirical dimension reduction techniques, such as factor analysis (FA) or principal component analysis (PCA), are performed on the covariance matrix of the returns of the risky assets. It gives rise to *eigenfactors* arising from an eigenvalue decomposition of the covariance matrix.

However, factors can be substituted by specific portfolios, especially if they represent different strategies that an investor can pursue at a low cost. The factors represent the various sources of risk present in the market to which an investor is exposed.

We now introduce the following linear factor model specification. The returns of a set of N risky assets indexed by $i = 1, \cdots, N$ are assumed to be wide-sense stationary, and can be expressed, for $t = 1, \cdots, T$, by:

$$\mathbf{r}_t = \boldsymbol{\alpha} + \mathbf{B}\,\mathbf{f}_t + \epsilon_t, \qquad [1.4]$$

where:

– $\mathbf{r}_t = [r_{1,t}, \cdots, r_{N,t}]'$ is the N-dimensional vector of the risky asset returns at time t;

– $\mathbf{f}_t = [f_{1,t}, \cdots, f_{K,t}]'$ is the K-dimensional vector of values of *common risk factors* at t whose covariance matrix is $\Sigma_{\mathbf{f}}$;

– $\mathbf{B} = [\mathbf{b}_1, \cdots, \mathbf{b}_K]$ is a $N \times K$ matrix where each element $b_{i,k} \in \mathbb{R}$ defines the exposure (or sensitivity) of the asset i to risk factor k. The sensitivities of asset i to the K factors are the K-dimensional row-vector $\mathbf{b}_i' = [b_{i,1}, \cdots, b_{i,K}]$. It is also referred to as the *beta*;

– $\boldsymbol{\alpha} = [\alpha_1, \cdots, \alpha_N]'$ denotes the N-dimensional vector of intercepts and is called *alpha*; and

– $\boldsymbol{\epsilon}_t = [\epsilon_{1,t}, \cdots, \epsilon_{N,t}]'$ is the N-dimensional vector of the zero-mean asset-specific residual returns whose covariance matrix is $\boldsymbol{\Sigma}_\epsilon$.

In [1.4], the unknown $N \times (K+1)$ matrix of parameters is $\boldsymbol{\Theta} = [\boldsymbol{\theta}_1, \cdots, \boldsymbol{\theta}_N]' = \left[\boldsymbol{\alpha}\ \mathbf{B}\right]$ and the inputs of the model are $\mathbf{r}_t$ and $\mathbf{f}_t$. Let $\boldsymbol{\epsilon}_i = [\epsilon_{i,1}, \cdots, \epsilon_{i,T}]'$ and $\mathbf{f}_k = [f_{k,1}, \cdots, f_{k,T}]'$ denote the T-dimensional vectors of, respectively, the residual returns of asset i and the values of factor k.

Additional assumptions are made for [1.4]:

A1) The residual returns are uncorrelated with each of the factors: $\mathbb{E}(\boldsymbol{\epsilon}_i \mathbf{f}_k') = \mathbf{0}_T$, $i = 1, \cdots, N$, $k = 1, \cdots, K$.

A2) The residual returns are temporally uncorrelated: $\mathbb{E}(\boldsymbol{\epsilon}_{t_1} \boldsymbol{\epsilon}_{t_2}') = \mathbf{0}_N$ for $t_1 \neq t_2$.

A3) The residual returns are uncorrelated, that is $\boldsymbol{\Sigma}_\epsilon$ is a diagonal matrix.

Assumption A3 means that the only sources of correlation among asset returns are those that arise from their exposures to the factors and the covariances among the factors. Residuals of asset returns are assumed to be unrelated to each other and hence totally *specific* to each asset. In other words, the risk associated with the residual return is *idiosyncratic* to the asset in question.

Note that the covariance matrix of the asset returns specified by [1.4] is:

$$\mathbf{\Sigma_r} = \mathbf{B}\,\mathbf{\Sigma_f}\,\mathbf{B}' + \mathbf{\Sigma}_\epsilon. \qquad [1.5]$$

The right-hand side of [1.5] consists of two distinct terms: $\mathbf{B}\,\mathbf{\Sigma_f}\,\mathbf{B}'$ is called the *systematic risk*, that is the risk explained by the K common factors and also known as *non-diversifiable risk*, *beta risk* or *market risk*.

The second term, $\mathbf{\Sigma}_\epsilon$, is called the *idiosyncratic risk*, also known as *diversifiable risk* or *asset-specific risk* and is totally specific to the assets.

As opposed to the *non-diversifiable systematic risk*, an investor has the possibility to reduce the amount of the *asset-specific risk* by properly diversifying his (or her) investments. Note also that $\mathbf{\Sigma_f}$ is diagonal when *eigenfactors* are selected.

When N is large, using [1.5] helps to narrow down the dimensionality to estimate the covariance matrix $\mathbf{\Sigma_r}$. On the one hand, the number of terms that must be estimated reduces significantly and on the other hand, $K < T$ usually meets the requirement to obtain an invertible estimated covariance matrix.

1.3. Why factor models in finance?

1.3.1. *Style analysis*

Let us consider a portfolio of n risky assets whose returns are $r_{i,t}, i = 1, ..., N$. The portfolio returns are denoted by $r_{p,t}$ and are obtained from $r_{i,t}, i = 1, ..., N$ by the following formula:

$$r_{p,t} = \sum_{i=1}^{N} a_i r_{i,t} = \mathbf{a}'\mathbf{r}_t,$$

where $\mathbf{r}_t = [r_{1,t}, \cdots, r_{N,t}]'$ and $\mathbf{a} = [a_1, \cdots, a_N]'$ is the vector of (fixed) weights of assets in the portfolio. The distribution of $r_{p,t}$ depends directly on the joint distribution of the vector $\mathbf{r}_t$, which can be considered as the factors explaining the portfolio performance. However, these factors can be highly correlated and/or unobserved (when we do not know the portfolio manager investment universe). A more parsimonious and tractable representation is then obtained using a small number of observed factors correlated with the portfolio performance. However, we must consider in this case the potential error made in explaining portfolio returns with a set of "common" factors. This explains why we introduce error terms in the linear factor representation.

A given portfolio allocation basically reflects the portfolio manager's bets. If we assume that these bets remain unchanged over the whole observation period, then the approach described above is relevant. However, from a practical standpoint, these bets are in general time-varying. The portfolio manager reallocates his/her portfolio on a continuous basis, and only the average exposure to factors is obtained through this classical return-based style analysis (see [SHA 92] for greater details on this approach). The linear combination of factors exposures and their respective performance gives then the strategic or long-term portfolio benchmark. The value added by the portfolio manager (or market timing ability) is then defined as the return difference between this portfolio returns and the strategic long-term benchmark (see [DAR 12]).

As a portfolio manager's bets are made on a continuous basis, it could also be interesting to track their impacts on the portfolio performance in the short term. Tactical portfolio allocation decisions rely on short-term portfolio manager's forecasts of risk premia (bets on factors) and can, as a result, also be captured using a linear factor model, but with time-varying exposures. Since information arrives randomly, and tactical bets are assumed to be responses to new information, we expect the exposure to risk factors to evolve randomly over time. Equation [1.4] in this case becomes:

$$\mathbf{r}_t = \boldsymbol{\alpha}_t + \mathbf{B}_t\,\mathbf{f}_t + \boldsymbol{\epsilon}_t,$$

where $\mathbf{B}_t$ (respectively, $\boldsymbol{\alpha}_t$) denotes the time-varying exposure of assets to factors (respectively, alpha). Risk factor exposures are not directly observed and must be filtered from $\mathbf{r}_t$.

Let us consider the case of a portfolio of hedge funds (or fund of hedge funds). Hedge funds are used in the following chapters to give empirical illustrations. Although the trend is very clearly toward more transparency, investors do not systematically have access to the full composition of hedge funds, and their evolution over time. Fund of hedge fund managers themselves do not always have a complete view of the risk factor exposures of their underlying investments, and, as a result, of the bets they implicitly make. This is all the more true when the trading frequency of the underlying funds is significantly higher than their reporting frequency (i.e. embedded risks can be dramatically different from those shown at a specific date), or when the number and the diversity of positions make it difficult to come up with accurate aggregated factor exposures. Tactical bets explicitly (at the portfolio level) and implicitly (at the underlying level) made by the fund of hedge fund manager can alternatively add up or cancel each other. Using a return-based style analysis therefore allows us to mitigate one of the main shortcomings of holding-based approaches, by capturing and assessing both effects concomitantly.

This example shows that this return-based style analysis must be done with time-varying parameter to filter from portfolio returns both long-term and short-term bets.

1.3.2. *Optimal portfolio allocation*

Let us consider a portfolio of N risky assets whose returns are $r_{i,t}, i = 1, ..., N$, with N large. According to Markowitz's [MAR 52] theory, optimal diversified portfolios are obtained by inverting the covariance matrix of asset returns $\mathbf{\Sigma_r}$. This matrix is unknown, and we have to estimate it from observations $r_{i,t}, i = 1, ..., N, t = 1, ..., T$ by using, for example, the empirical covariance matrix:

$$\hat{\mathbf{\Sigma}}_{\mathbf{r},\mathbf{T}} = \frac{1}{T}\sum_{t=1}^{T}(\mathbf{r}_t - \mathbb{E}(\mathbf{r}_t))(\mathbf{r}_t - \mathbb{E}(\mathbf{r}_t))',$$

where $\mathbf{r}_t = [r_{1,t}, \cdots, r_{N,t}]'$ is the vector of asset returns. If the dimension T is lower than N, then $\hat{\mathbf{\Sigma}}_{\mathbf{r},\mathbf{T}}$ is not invertible, and the optimal Markovitz's portfolios cannot be computed. Moreover, even in the case where N is lower than T, this empirical covariance matrix can have a determinant close to zero, and we can encounter numerical problems when trying to invert it with the usual algorithms. This is, in particular, the case when correlations between assets are high. Factor models can be used to manage these high correlations and therefore give an alternative to the direct numerical inversion approach.

Let us develop these computations in the very simple case [1.4]. We assume in the following that the risk-free rate is set to zero and excess risky asset returns $\mathbf{r}_t$ satisfy the following linear single risk factor model:

$$r_{i,t} = b_i f_t + \epsilon_{i,t}, i = 1, ..., N,$$

where the unobserved single factor f_t is assumed to be Gaussian $N(m_f, \sigma_f^2)$ and $\epsilon_{i,t}, i = 1, ..., N$ have Gaussian distribution $N(0, \sigma_\epsilon^2)$. The expected excess return is $\mathbb{E}(r_{i,t}) = b_i m_f$ and the idiosyncratic risk σ_ϵ^2. The systematic source of risk (i.e. the unobserved single factor) creates an additional individual risk equal to $b_i^2\sigma_f^2$, but also a covariance between excess returns of two different assets:

$$Cov(r_{i,t}, r_{j,t}) = b_i b_j \sigma_f^2,$$

for $i \neq j$. This covariance term can lead to correlation close to 1 when the systematic risk is high relative to the idiosyncratic risk. However, we can compute explicitly the mean-variance optimal allocation. The vector of efficient allocations in the N risky assets is indeed proportional to:

$$\mathbf{a}^* = V(\mathbf{r}_t)^{-1}\mathbb{E}(\mathbf{r}_t),$$

where $\mathbf{r}_t = [r_{1,t}, \cdots, r_{N,t}]'$. Using the factor structure, we can explicit the two first moments of $\mathbf{r}_t$ and get the following closed-form formula:

$$\mathbf{a}^* = (\sigma_\epsilon^2\,\mathbf{I} + \sigma_f^2\,\mathbf{B}\,\mathbf{B}')^{-1}\, m_f\,\mathbf{B} = \frac{m_f}{\sigma_\epsilon^2 + \sigma_f^2\,\mathbf{B}'\,\mathbf{B}}\,\mathbf{B},$$

where $\mathbf{B} = [b_1, \cdots, b_N]'$. We then obtain the efficient allocation by estimating the parameters involved in the previous formula, without having to numerically invert the covariance matrix.

In this example, the single factor is unobserved and then must be filtered from asset returns, or replaced by a proxy that is able to explain all the correlation structure observed between risky assets.

1.4. How to build factor models?

1.4.1. *Factor selection*

The factor selection problem is not new in the financial literature. The main issue is related to the delicate balance between using too many or too few factors. On the one hand, adding too many factors lowers the regressors efficiency when we estimate factor exposures using equation [1.4]. On the other hand, working with too few factors also has an important risk of missing the correlation structure observed between asset returns. This raises the question whether it is possible to build a factor selection methodology allowing us to consider only the appropriate factors.

The first solution consists of using a predefined set of observable factors, already documented in the financial literature for their ability to explain the cross-section of asset returns. These factors are used to build long- and short-term benchmarks in the style analysis of section 1.3.1. They are also used to reduce the dimension of the covariance matrix, when optimal portfolio allocation must be computed for a large set of risky assets in section 1.3.2. If we miss some risk factors in the style analysis, both long- and short-term benchmarks are misspecified, and the portfolio manager's added value is not correctly calculated. In particular, we can interpret positive risk-adjusted returns as a manager's skills and omit an important risk exposure. In the portfolio allocation example, omitted factors imply residual correlations between idiosyncratic risk terms. As these correlations are not taken into account in the calculation of the optimal portfolio, we underestimate the portfolio risk or, in other words, we overestimate the diversification effect.

A second solution is to use statistical approaches to filter from the asset returns distribution (and, in particular, the covariance matrix) unobservable factors that are able to explain the cross-section of asset returns. If this approach seems appealing to a statistical point of view, it has also many drawbacks from a financial point of view. First, it is, in general, difficult to give an economic interpretation of these statistical factors. Second, the factor representation is not unique, and any linear combination of a given set of factors defines an equivalent factor model. Third, the factor decomposition is, in general, time-varying and then difficult to interpret. Both approaches are described in detail in Chapter 2.

1.4.2. *Parameters estimation*

Once a set of factors is chosen, parameters estimation consists of computing numerical values of factors exposures. When parameters are not time-varying, least squares (LS) approaches can be used to compute these estimators, depending on the residual properties. However, both in the style analysis of section 1.3.1 and the optimal portfolio allocation problem of section 1.3.2, parameters can be time-varying and more complex statistical filtering approaches must be used (see e.g. [BOL 09, PAT 13]) such as the Flexible Least Squares [KAL 89, MAR 04], the Kalman filter (KF) [RON 08a, RON 08b, RAC 10] or Markov switching regimes [BIL 12].

The LS estimator can always be computed on rolling windows, updated each time we get new data. This provides a time series of estimators that have then a time-varying behavior. However, this approach, if useful when changes in parameters over time are smooth, can be misleading when big changes occur. A style analysis computed using rolling window LS can miss a sudden style rotation decided by the portfolio manager. On the contrary, this estimation method provides very good results when style drifts are small and/or implemented step by step.

The KF approach is by definition more reactive and can capture in theory quick style rotations. However, this approach works well only when the relevant factors are used in equation [1.4]. Indeed, assumptions made up of the statistical behavior of the residual returns are more constraining, and any failure can have a huge impact of the filtered time-varying exposures. It is, for example, the case when residual returns do not follow a Gaussian distribution. Commonalities and discrepancies between LS and KF are discussed in Chapter 3, and an enhanced version of KF is discussed in Chapter 4.

1.5. Historical perspective

Factor models have been the focus of numerous studies in empirical finance since Treynor [TRE 62], Sharpe [SHA 64], Lintner [LIN 65] and Mossin [MOS 66] developed the capital asset pricing model (CAPM) in the 1960s.

1.5.1. *CAPM and Sharpe's market model*

CAPM lays the foundation for all the existing factor models. It gives the theoretical *equilibrium* relationship that should occur between the returns and the risks of individual assets with regard to the market returns.

Its conception is based on two fundamental financial concepts: market equilibrium that occurs if the amount of demand is balanced by the amount of supply (represented by the market portfolio), and, as a result, the mean-variance efficiency of the market portfolio.

Given the risk-free rate r_f and the risk σ_m of the market portfolio m, the CAPM stipulates that the returns we expect from individual assets in excess of r_f are given in proportion to the excess returns expected from the market, as follows:

$$\mathbb{E}(r_{i,t}) - r_f = \beta_i \left(\mathbb{E}(r_{m,t}) - r_f\right), \qquad [1.6]$$

where $\beta_i = \gamma_{im}/\sigma_m^2$ is related to the amount of risk given by i, with γ_{im} being the covariance between r_i and r_m. Larger values of β_i correspond to larger expected return and larger risk for asset i. The term $(\mathbb{E}(r_{m,t}) - r_f)$ is called market risk premium.

In practice, the unknown parameter β_i is estimated through the following univariate regression using historical data for the asset returns, the market portfolio and the risk-free rate:

$$r_{i,t} - r_f = \alpha_i + \beta_i (r_{m,t} - r_f) + \epsilon_{i,t}, \qquad [1.7]$$

where $\epsilon_{i,t}$ satisfies $\mathbb{E}(\epsilon_{i,t}) = 0$. Additional parameter α_i is the asset *alpha*. The strict form of [1.6] specifies that *alpha* must be zero and that deviation from zero is the result of temporary disequilibrium. With $\alpha_i = 0$ in [1.7], this model coincides with Sharpe's market model that is an output of [1.6]. The market portfolio is usually replaced by a major standard equity index (such as the S&P500) since Black [BLA 72] has shown that in market equilibrium, such market-value weighted indices are always efficient. The parameter α_i can be used to perform fitted performance analysis (see [DAR 10]).

CAPM results can be represented by the so-called security market line (SML) as shown in Figure 1.2. The SML is the plot of the expected return of any asset i as a function of its *beta*, as given by [1.6]. It is obtained for a fixed period of time through [1.6] where r_f is the mean of the risk-free asset returns, $\mathbb{E}(r_{m,t})$ is the mean of the returns for the proxy of the market portfolio and where some theoretical values for the β_is are chosen. For each risky asset i, α_i and β_i are estimated through [1.7] and $\hat{\mu}_i$ is the estimated mean of its returns. According to CAPM, given $\hat{\beta}_i$, if $\hat{\mu}_i$ does not lie on the theoretical SML, then the asset is mispriced.

In practice, this model has received several criticisms (e.g. see Roll's critique [ROL 77]). First, the market portfolio is unobservable. Standard equity indices usually substitute for it but do not reflect all the wealth in the economy as the market portfolio does. Second, the mean-variance approach depends only on the first two moments of the asset returns, which is too restrictive.

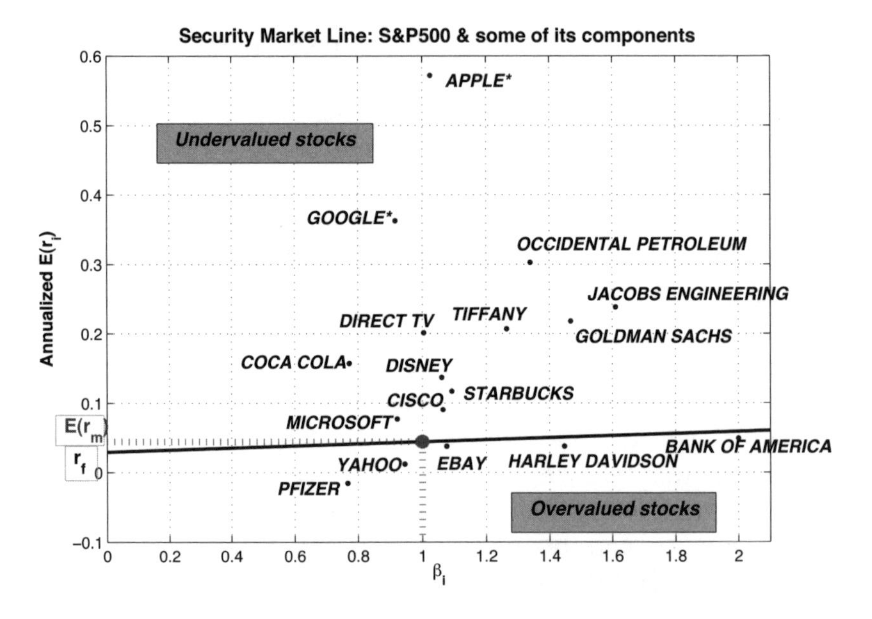

Figure 1.2. *The security market line (SML, the black line) shows the expected return of asset i as a function of its beta, as given by* [1.6], *and given in annualized values over the period August 2004 / October 2010:* $r_f = 2.93\%$, $\mathbb{E}(r_{m,t}) = 4.41\%$, *and* $\sigma_m = 22.81\%$. *Here,* r_f *is the average rate of the US T-Bill 1 year over the whole period,* m *is the S&P500 and the risky assets are some of its components. Each (blue) point has for coordinates* $(\hat{\beta}_i, \hat{\mu}_i)$ *– the estimated values, for asset* i, *of* β_i *and* $\mathbb{E}(r_{i,t})$. *If* $\hat{\mu}_i$ *is lower (respectively, higher) than* $\mathbb{E}(r_{i,t})$, *obtained from* [1.6] *given* r_f, $\mathbb{E}(r_{m,t})$ *and* $\hat{\beta}_i$, *then the asset is said to be under (respectively, over) valued. Stocks with an * have significant (at 99.5%) non-zero* α_i

Finally, empirical studies often invalidate the CAPM because of its strong assumptions and show that more than one factor is necessary to identify the market risks.

1.5.2. *APT for arbitrage pricing theory*

Ross [ROS 76] and Roll and Ross [ROL 77, ROL 80] developed an alternative *arbitrage* model to the CAPM, called APT model. APT follows from two basic postulates:

P1) the risky assets follow a K-*factor structure* like [1.4] with three additional assumptions:

A4) $K << N$;

A5) $\mathbb{E}(f_k) = 0$, $k = 1, \cdots, K$ that leads to $\mathbb{E}(\mathbf{r}_t) = \boldsymbol{\alpha}$;

A6) $\exists\, s < \infty$ such that $\sigma^2_{\epsilon_i} = \mathbb{E}(\epsilon^2_{i,t}) \leq s^2$.

P2) pure *arbitrage* profits are impossible.

Pure arbitrage profits are risk-free profits at zero cost: an investor can earn a positive return on any combination of assets without undertaking risk and without making some net investment of funds. With P1 and P2, APT stipulates that the risk premiums of the risky assets are given through a linear combination of the factor *risk prices* weighted by the factor sensitivities $\mathbf{b}_i$ of [1.4]:

$$\mathbb{E}(r_{i,t}) = r_f + \sum_{k=1}^{K} b_{i,k}\, \lambda_k, \qquad [1.8]$$

where λ_k defines the *price of risk* for factor k. The full APT expression is then obtained by replacing α_i, $i = 1, \cdots, N$ in [1.4] by [1.8]. Jointly estimating the K-dimensional vectors $\mathbf{b}_i$ and $\boldsymbol{\lambda} = [\lambda_1, \cdots, \lambda_K]'$ is not trivial. In practice, a two-pass methodology can be used: first estimate $\mathbf{b}_i$ through [1.4] (giving rise to $\hat{\mathbf{b}}_i$), then use $\hat{\mathbf{b}}_i$ as input to estimate $\boldsymbol{\lambda}$ through:

$$r_{i,t} = r_f + \sum_{k=1}^{K} \hat{b}_{i,k}\, \lambda_k + u_{i,t},$$

where $u_{i,t}$ represents zero-mean residual errors with usual assumptions. Unlike the CAPM, APT does not provide any information about the nature of the K factors.

1.6. Glossary

Volatility

The volatility σ of an asset a refers to the standard deviation of the continuously compounded returns of a within a specific time horizon. Volatility is usually expressed in annualized terms: if the volatility σ is computed using daily (respectively, weekly) returns and if we consider that a year is made us of 250 business days (respectively, 52 weeks), then the annualized volatility will be equal to $\sqrt{250}\,\sigma$ (respectively, $\sqrt{52}\,\sigma$).

Risk

The risk of holding an asset a may be quantified by the volatility of a.

Risk-free rate

The theoretical rate of return of an investment with zero risk; the risk-free rate, commonly denoted r_f, represents the minimum return an investor expects for any investment when taking no risk. This is often considered as a lower bound of what should at least give any riskier investment since bearing any risk should be more remunerating than a risk-free investment.

Expected (excess) return

The expected return of an asset a is the return on a expected in the future and is computed by $\mathbb{E}(r_a)$. The expected excess return is the expected return on a in excess of the return given by the risk-free rate (or by any other market measure): $\mathbb{E}(r_a) - r_f$.

Risk premium

The risk premium of an asset a is the expected return on a in excess of the risk-free rate, that is $\mathbb{E}(r_a) - r_f$.

Zero-cost portfolio

A zero-cost portfolio is a portfolio for which the weights add up to zero. For example, the excess return of an asset a, that is $\mathbb{E}(r_a) - r_f$ is a zero-cost portfolio. It invests 100% in a and -100% in the risk-free rate. The amount borrowed at r_f is invested in a. As the investor will have to pay the interest for borrowing money at r_f, he/she would therefore expect to receive more than r_f in return for his/her investment in a.

Tradable portfolio

A portfolio is said to be tradable if it is a zero-cost portfolio.

Portfolio diversification

Diversifying a portfolio is the act of adding more investments to one's portfolio to reduce the risk inherent in any one investment. It increases the possibility of making a profit, or at least avoiding a loss. In general, the broader the diversification, the less is the risk and the return. Adding more investments to the portfolio for diversification involves subdividing the portfolio among many smaller investments. If the portfolio's size increases instead of remaining constant, the portfolio's risk may not decrease especially if the assets added in the portfolio are uncorrelated. Here is a very simple example of diversification benefits on the variance of the portfolio. If the N assets in the portfolio are mutually uncorrelated and have identical variances σ^2, portfolio variance is minimized by holding all assets in the equal proportions $1/N$. Then, the portfolio variance equals σ^2/N that is monotonically decreasing in N. So, even if the added assets are uncorrelated, the portfolio variance decreases. Benefits of diversification amplify when adding negatively correlated assets in the portfolio. The modern understanding of diversification dates back to the work of Harry Markowitz [MAR 52] in the 1950s.

Market portfolio

A market portfolio is a perfectly well-diversified portfolio and represents the evolution of the market as a whole. In the factor model framework, a perfectly diversified portfolio admits a pure factor structure, that is a factor structure where there is no additional idiosyncratic risk.

Market efficiency

The market is said to be efficient if the price of the assets in the market reflects all information available.

Market equilibrium

The market equilibrium occurs if the amount of demand is balanced by the amount of supply. In this condition, the prices should tend not to change unless demand or supply change.

Arbitrage

Arbitrage is the possibility of a risk-free profit at zero cost. For example, if the same asset does not trade at the same price on all markets, buying and selling this asset simultaneously and instantaneously on two different markets will take advantage of the price difference.

Market capitalization

Market capitalization, also known as Market Cap. or cap. (MC), is the number of outstanding shares of a firm times the price of its share in the market. MC measures the size of the firm. It changes every day because of the quoted firm price (P). For example, in October 24, 2010, the six largest MC in the world were (T$ is for trillions US dollars and B$ for billions US dollars):

(1) Telecom Brasil with MC = T$6.57 and P = $5.99,

(2) Exxon Mobil with MC = B$337.69 and P = $66.32,

(3) Apple with MC = B$282.77 and P = $309.52,

(4) PetroChina with MC = B$229.20 and P = $125.23,

(5) BHP Billiton with MC = B$224.83 and P = $80.81, and

(6) Microsoft with MC = B$219.97 and P = $25.42.

Chapter 2

Factor Selection

2.1. Introduction

The use of factor models requires the determination of the nature and the number of the factors. Two approaches can be distinguished in the literature: the empirical *ad hoc* approach and the statistical approach which involve subspace methods and lead to the so-called *eigenfactors*.

Empirical studies generally compare the two approaches (see [CHE 93, CON 95, CHA 98, DAR 11]) and mostly emphasize the superiority of eigenfactors over macro-economic factors regarding their expected properties (uncorrelated factors) and the percentage of the explained part of the systematic returns they contain. Eigenfactors are nevertheless difficult to relate to observed macro-economic factors which lower in a certain sense their attractiveness to practitioners.

This chapter is organized as follows: section 2.2 focuses on the empirical *ad hoc* approach and presents three reference models that are widely used in the literature. These models are all based on the factor representation described in Chapter 1 (see section 1.4), but highlight the nature (and the number) of the factors to be used to explain specific asset class returns. In section 2.3, we present the statistical approach. As opposed to the empirical factor selection where *a priori* knowledge on the observed data is required, the purely statistical approach extracts a smaller group of implicit eigenfactors from the time series of the observed returns which explain a large amount of information contained in the observed data. The problem is then to determine the order of the resulting model. Some classical techniques, arising from the information theory, are described in section 2.4. Sections 2.5–2.7 are complementary sections giving some light on related problems to this approach such as the estimation of the covariance matrix of the data, the similarity of the approach with subspace methods and the extension of this approach to large panel data.

2.2. Qualitative know-how

In an empirical *ad hoc* approach, the nature and the number of factors are chosen with some *a priori* consideration of the market conditions. The factors are observable market risk factors and are included in the model through a discretionary choice. They also depend on the class of assets under study so that the mis-specification risks are non-negligible.

A substantial amount of work is available on the subject in the literature, so, in this section, we present three major reference models: the factor model of Fama and French [FAM 93], the Chen *et al.* model [CHE 86] – two models developed to explain the risks in the US equity market, and the risk-based factor model of Fung and Hsieh [FUN 01, HSI], a model adapted to explain the risks bourne by the trend-follower hedge funds.

2.2.1. *Fama and French model*

The three-factor model of Fama and French [FAM 93] is an extension to the capital asset pricing model (CAPM). Observing that *small caps* and *value* stocks outperformed the market on a regular basis, Fama and French included two additional factors to Sharpe's market model: *SMB* for *small minus big* (market capitalization, MC), which reflects the relative performance of the small versus large caps stocks, and *HML* for *high minus low* (book-to-market value), which reflects the relative performance of the high-valued versus the low-valued stocks. MC gives information about the size of the firm (equal to the stock price times the number of its outstanding shares). Book-to-market ratio is related to the accounting value of the firm (equal to the MC of the firm divided by the accounting value of all the firm's shares).

For example, SMB factor may be constructed at time t as follows, for a fixed period of time of length T (from $t-T$ to t):

– consider a large universe of stocks in the equity market, for example take the constituents of the Russel 2000 equity Index at t;

– for each stock in the universe, get its MC from $t - T$ to t;

– for each date, rank the stocks according to their relative MC;

– to obtain the relative performance of the small versus the large caps stocks, compute, for each date, an equi-weighted (or a market cap weighted) portfolio composed of the stocks having an MC lower (respectively bigger) than a specific value;

– SMB factor returns are then computed as the difference between the returns of the small MC portfolio and the big MC portfolio.

The procedure is identical to construction of the HML factor but is based on the values of the book-to-market value of the stocks instead of their MC values.

This approach is used, for example, by Morningstar to rate stocks and mutual funds using a five stars grid[1].

The four-factor model of Carhart [CAR 97] is an extension of the three-factor model of Fama and French [FAM 93], including a momentum factor. The model is shown to explain persistence in equity mutual funds' mean and risk-adjusted returns.

2.2.2. *The Chen et al. model*

The Chen *et al.* [CHE 86] model is a particular case in the arbitrage pricing theory (APT) framework. Studying the stocks in the US market, the authors identified a set of economic state variables that were significant in explaining the expected stock returns: industrial production, changes in the risk premium, twists in the yield curve, measures of unanticipated inflation and changes in inflation during high volatile periods for these variables. Whereas a value-weighted stock equity index explains a significant portion of the time series variability of stock returns, it has an insignificant influence on pricing (i.e. on expected return) when compared against the above state variables.

As opposed to the Fama and French model, this factor model contains only observed macro-economic factors, so that there is no other specific procedure to construct such a model than getting from a data provider the time series of the above cited factors on the period where the study is conducted.

1 See http://www.morningstar.com

2.2.3. *The risk-based factor model of Fung and Hsieh*

The eight-factor model of Fung and Hsieh (*F&H model* [FUN 01, HSI]) is a risk factor model adapted to trend-follower hedge funds. Hedge funds are non-quoted speculative funds seeking high expected returns theoretically uncorrelated with the market. Hedge fund managers can borrow money to substantially *leverage* their profitability, often at a factor of 10 to 1, or more. They usually purchase derivative products (such as options) and take *short* positions to also benefit from downward movements of the assets. The positions are managed dynamically. As a result, hedge funds exhibit nonlinear option-like exposures to the standard asset classes that cannot be captured by the traditional linear factor models. A growing body of literature is devoted to finding accurate hedge fund risk factors (see [AGA 00a, AGA 00b, AGA 04, AGA 09, SAD 10]), and one of the most famous hedge fund risk models is the F & H model. It recommends seven asset-based risk factors and one emerging market risk factor to describe the risks of the *trend-follower* hedge funds and is composed of [HSI]:

1) three customized *trend-following* risk factors, which are specific portfolios, constructed by the authors using an appropriate mix of options (or more precisely *lookback straddles*) that allow the characterization of pure directional positions taken by the *trend-follower* hedge fund managers on three different asset classes:

i) *PTFSCOM*, representing the returns of a specific portfolio invested in commodities,

ii) *PTFSBD*, representing the returns of a specific portfolio invested in bonds, and

iii) *PTFSFX*, representing the returns of a specific portfolio invested in foreign exchange;

2) two *equity-oriented* risk factors, which are based on equity market indices:

i) *SP500* represents the total return of the Standard and Poors 500 equity index, and

ii) *EquitySizeSpread* represents the performance of the smallest MC of the US Equities as being the difference between the total returns of the large Russel 2000 equity index (composed of the 2,000 largest MC in the US equity market) and *SP500*;

3) two *bond-oriented* risk factors, representing the bond market:

i) *Bond10y* is the monthly change in the 10-year treasury constant maturity yield, and

ii) *CreditSpread* is the monthly change in the Moody's Baa yield less 10-year treasury constant maturity yield;

4) the recently added emerging market index factor:

i) *MXEF*, representing the total returns of the Morgan Stanley Capital International (MSCI) emerging market index.

The eight risk factors of Fung and Hsieh are available on the author's webpage (see [HSI]) and are quoted monthly from January 1994. Table 2.1 gives some basic statistics on these factors on the period January 1994 to December 2012. These statistics are: the annualized return, the annualized volatility, the ratio between annualized return and the annualized volatility (denoted as the Sharpe ratio), the minimum and the maximum monthly returns on the period, and the skewness and the kurtosis of the return distribution.

Name	Annualized returns %	Annualized volatility %	Sharpe ratio	Minimum return %	Maximum return %	Skewness	Kurtosis
SP500	7.1	15.44	0.46	−16.94	10.77	−0.66	3.96
EquitySizeSpread	1.19	11.70	0.10	−16.36	18.43	0.27	7.92
MXEF	6.59	24.26	0.27	−29.29	16.66	−0.72	4.74
Bond10y	−3.18	24.67	−0.13	−26.93	27.56	0.17	5.61
PTFSCOM	−6.98	47.49	−0.15	−24.65	64.75	1.12	5.11
PTFSBD	−18.66	53.79	−0.35	−26.63	68.86	1.38	5.49
PTFSFX	−7.13	67.86	−0.10	−30.13	90.27	1.35	5.50
CreditSpread	5.27	26.12	−0.20	−15.55	38.25	1.09	6.24

Table 2.1. *Averaged statistics and correlation for the eight risk factors of Fung and Hsieh. Computation of the statistics is made on monthly returns from January 1994 to December 2012*

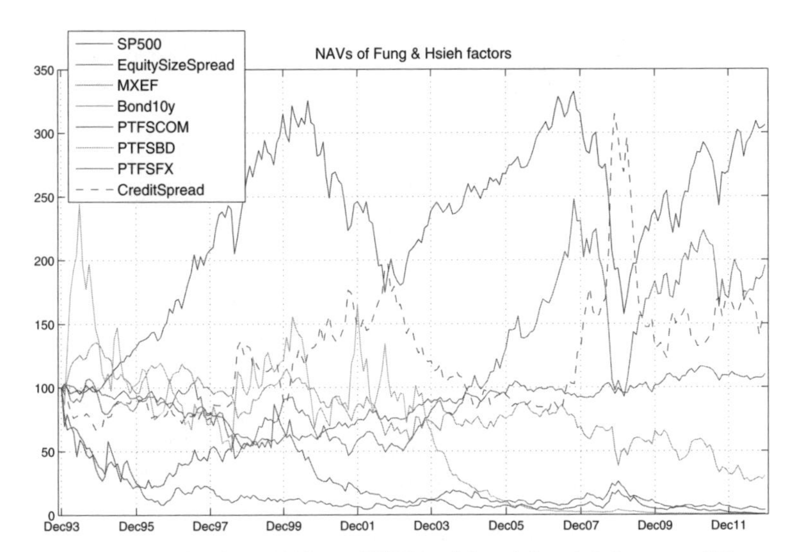

Figure 2.1. *Net Asset Values (NAVs) of the eight risk factors of Fung and Hsieh. The NAVs are based at 100 in December 1993 and end in December 2012*

As an illustration, Figure 2.1 shows the evolution of these factors if we had invested 100 in each of them in December 1993, and Figure 2.2 represents the correlation matrix between these factors computed with monthly factor returns from January 1994 to December 2012. Two groups of anti-correlated factors can be distinguished: the first group, composed of *SP500*, *EquitySizeSpread*, *MXEF* and *Bond10y* are negatively correlated with the other four factors. Moreover, within each group, we observe a quite weak (positive or negative) correlation between the factors, excepted for the two equity-based risk factors (which are positively correlated at approx. 60%) and the two bond-related risk factors (which present a negative correlation at approx. 45%).

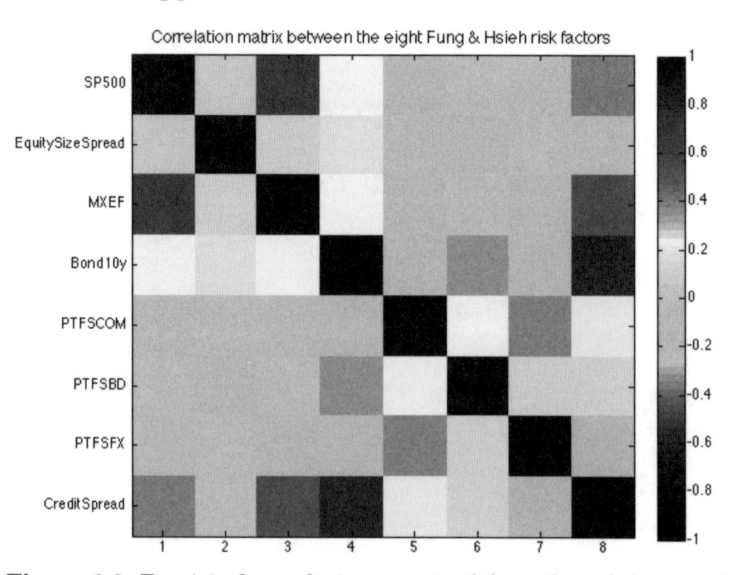

Figure 2.2. *Empirical correlation matrix of the eight risk factors of Fung and Hsieh from December 1993 and end in December 2012 (see color plate section)*

Macro-economic factors are naturally correlated. To avoid correlation between the returns of the different asset class factors, the factors may be selected through a stepwise regression technique [AGA 00b]: the factors are entered or removed from the model, one at a time, depending on their relative discriminant power. Used in its ascending version, the single best factor is chosen first, then paired with each of the other variables, and the second best factor is chosen, and so on. The order of the model increases until adding a new factor does not improve the discriminant power of the previous model. Used in its descending version, the initial model contains all the pre-selected factors and removes, one at a time, the worst factor, until removing one of the remaining factors does not improve the discriminant power of the previous model.

2.3. Quantitative methods based on eigenfactors

In this section, we denote by *eigenfactors* the factors obtained from the observations using the eigenvector decomposition of the covariance matrix of the returns. In the literature on empirical quantitative finance, *eigenfactors* might be viewed as special *latent factors* that are a subset of vectors obtained from linear combinations of returns. Such methods are often used when there is no *a priori* knowledge on the economic factors that may drive the data under study and give the remarkable property of yielding to orthogonal factors especially when subspace methods are used to derive such factors.

In the statistical literature, the most frequently used methods are the Factor Analysis method and the Principal Component Analysis (PCA) method. Mainly used in order to reduce the dimensionality of the matrix of observations, such methods use the eigenvector decomposition of the correlation matrix and yield to components that are constructed as linear combinations of returns. The most important result is that the original data may be approximated with a high percentage of accuracy by keeping only a few eigenvectors that are selected such that their associated eigenvalues are the highest. The difficulty is therefore to determine the selection threshold. This part of the problem is discussed in section 2.4.

2.3.1. *Notation*

Let us assume that returns are stored in a $T \times N$ matrix $\mathbf{R}$ whose N columns represent T consecutive asset returns sampled at a regular frequency (e.g. on a daily basis, ...):

$$\mathbf{R} = \left[\mathbf{r}_1 \, \mathbf{r}_2 \cdots \mathbf{r}_N\right] = \begin{bmatrix} r_{1,1} & r_{2,1} & \cdots & r_{N,1} \\ r_{1,2} & r_{2,2} & \cdots & r_{N,2} \\ \vdots & \vdots & r_{j,t} & \vdots \\ r_{1,T} & r_{2,T} & \cdots & r_{N,T} \end{bmatrix}. \qquad [2.1]$$

If $\boldsymbol{\mu} = \mathbb{E}(\mathbf{R}) = \left[\mu_1 \, \mu_2 \cdots \mu_N\right]'$ denotes the N-dimensional mean vector, then:

$$\bar{\mathbf{R}} = \left[\mathbf{r}_1 - \mu_1 \; \mathbf{r}_2 - \mu_2 \cdots \mathbf{r}_N - \mu_N\right], \qquad [2.2]$$

is the centered version of $\mathbf{R}$.

If $\boldsymbol{\sigma} = \left[\sigma_1 \, \sigma_2 \cdots \sigma_N\right]'$ denotes the N-dimensional standard deviation vector and $\mathbf{D}$ a diagonal $N \times N$ matrix whose principal diagonal is $\boldsymbol{\sigma}$, such that:

$$\mathbf{D} = \begin{bmatrix} \sigma_1 & & & 0 \\ & \sigma_2 & & \\ & & \ddots & \\ 0 & & & \sigma_N \end{bmatrix}, \qquad [2.3]$$

then let us define $\widetilde{\mathbf{r}}_j = \mathbf{D}^{-1}(\mathbf{r}_j - \mu_j) = \mathbf{D}^{-1}\bar{\mathbf{r}}_j$ as the standardized version of $\mathbf{r}_j$ which is then a zero-mean vector with a unitary variance. We also have:

$$\widetilde{\mathbf{R}} = \left[\widetilde{\mathbf{r}}_1\ \widetilde{\mathbf{r}}_2 \cdots \widetilde{\mathbf{r}}_N\right] = \left[\frac{\mathbf{r}_1 - \mu_1}{\sigma_1}\ \frac{\mathbf{r}_2 - \mu_2}{\sigma_2} \cdots \frac{\mathbf{r}_N - \mu_N}{\sigma_N}\right]$$
$$= \left[\frac{\bar{\mathbf{r}}_1}{\sigma_1}\ \frac{\bar{\mathbf{r}}_2}{\sigma_2} \cdots \frac{\bar{\mathbf{r}}_N}{\sigma_N}\right]. \qquad [2.4]$$

The covariance matrix $\Sigma_{\mathbf{R}}$ of R, defined as $\Sigma_{\mathbf{R}} = Var(\mathbf{R}) = \mathbb{E}[\bar{\mathbf{R}}' \bar{\mathbf{R}}] = \mathbb{E}(\mathbf{R}' \mathbf{R}) - \mu \mu'$, can therefore be expressed as:

$$\Sigma_{\mathbf{R}} = \begin{bmatrix} \sigma_1^2 & \sigma_{12} & \cdots & \sigma_{1N} \\ \sigma_{21} & \sigma_2^2 & \cdots & \sigma_{2N} \\ \vdots & \vdots & \ddots & \vdots \\ \sigma_{N1} & \sigma_{N2} & \cdots & \sigma_N^2 \end{bmatrix}. \qquad [2.5]$$

Let us note that $\Sigma_{\mathbf{R}}$ is symmetric. If $\Sigma_{\mathbf{R}}$ is full rank, then $\Sigma_{\mathbf{R}}$ is also definite positive, which means that its N eigenvalues are strictly positive. If it is not full rank then the matrix is semi-definite positive.

Several methods may be used to estimate the covariance matrix. The traditional sample covariance matrix (SCM) and robust approach with M estimators are presented in section 2.5.

2.3.2. *Subspace methods: the Principal Component Analysis*

The most commonly used method in financial applications to obtain *latent* factors is the PCA that makes an eigen-decomposition of the covariance matrix of returns. In this section, we assume that the covariance matrix is either known or estimated with any method (like those presented in section 2.5 or any other one), and we use the notation $\widehat{\Sigma}$ to designate the covariance matrix estimate.

Assume that the observations are standardized, that is we use $\widetilde{\mathbf{R}} = \left[\widetilde{\mathbf{r}}_1\, \widetilde{\mathbf{r}}_2 \cdots \widetilde{\mathbf{r}}_T\right]'$ as defined in [2.4] with $\widetilde{\mathbf{r}}_t = \mathbf{D}^{-1}(\mathbf{r}_t - \boldsymbol{\mu})$. In such a case, we then have $\mathrm{tr}(\boldsymbol{\Sigma}_{\widetilde{\mathbf{R}}}) = N$.

If $\widehat{\boldsymbol{\Sigma}}_{\widetilde{\mathbf{R}}}$ designates an estimate version of $\boldsymbol{\Sigma}_{\widetilde{\mathbf{R}}}$, then its eigen-decomposition can be written as:

$$\widehat{\boldsymbol{\Sigma}}_{\widetilde{\mathbf{R}}} = \sum_{k=1}^{N} \lambda_k\, \mathbf{u}_k\, \mathbf{u}_k' = \mathbf{U}\,\boldsymbol{\Gamma}\,\mathbf{U}', \tag{2.6}$$

where $\mathbf{U} = \left[\mathbf{u}_1\, \mathbf{u}_2 \cdots \mathbf{u}_N\right]$ is the $N \times N$ unitary matrix composed of the N eigenvectors corresponding to the N eigenvalues $\{\lambda_k\}_{k=1}^{N}$ stored in a decreasing order $(\lambda_1 \geq \lambda_2 \geq \cdots \geq \lambda_N)$ in the $N \times N$ diagonal matrix $\boldsymbol{\Gamma}$.

Let us say that the K largest eigenvalues are stored in $\boldsymbol{\Gamma}_K$. We then have:

$$\boldsymbol{\Gamma} = \begin{pmatrix} \boldsymbol{\Gamma}_K & \mathbf{0} \\ \mathbf{0} & \boldsymbol{\Gamma}_{N-K} \end{pmatrix}$$

$$\text{with } \boldsymbol{\Gamma}_K = \begin{pmatrix} \lambda_1 & & 0 \\ & \ddots & \\ 0 & & \lambda_K \end{pmatrix} \text{ and } \boldsymbol{\Gamma}_{N-K} = \begin{pmatrix} \lambda_{K+1} & & 0 \\ & \ddots & \\ 0 & & \lambda_N \end{pmatrix}, \tag{2.7}$$

and

$$\mathbf{U} = \left[\mathbf{U}_K\; \mathbf{U}_{N-K}\right], \tag{2.8}$$

so that equation [2.6] can easily be rewritten as:

$$\begin{aligned} \widehat{\boldsymbol{\Sigma}}_{\widetilde{\mathbf{R}}} &= \sum_{k=1}^{K} \lambda_k\, \mathbf{u}_k\, \mathbf{u}_k' + \sum_{k=K+1}^{N} \lambda_k\, \mathbf{u}_k\, \mathbf{u}_k' \\ &= \mathbf{U}_K\, \boldsymbol{\Gamma}_K\, \mathbf{U}_K' + \mathbf{U}_{N-K}\, \boldsymbol{\Gamma}_{N-K}\, \mathbf{U}_{N-K}'. \end{aligned} \tag{2.9}$$

In many applications, K is linked to the order of the model: it is assumed that the observations come from $K < N$ sources and the $N - K$ remaining sources may be considered as noise. As an example, MUSIC algorithm (MUltiple SIgnal Characterization) [STO 89, STO 97, TRE 02] is commonly used for determining the number of sources in an array antenna as well as the angle of arrival. The signal space, defined by the K largest eigenvalues, is orthogonal to the noise space (the $N-K$ smallest eigenvalues) so that projecting the observation on to the signal space allows for detecting the signal of interest.

Using the K-dimensional subspace defined by $\mathbf{\Gamma}_K$, and similarly by $\mathbf{U}_K$ the K eigenvectors corresponding to the K largest eigenvalues, then the K estimated eigen factors $\widehat{\mathbf{F}} = \left[\widehat{\mathbf{f}}_1 \cdots \widehat{\mathbf{f}}_K\right]$ are obtained by:

$$\widehat{\mathbf{F}} = \widetilde{\mathbf{R}}\,\mathbf{U}_K. \qquad [2.10]$$

By construction, $\widehat{\mathbf{f}}_k$ is a T-dimensional vector with:

$$\widehat{\mathbf{f}}_t = \mathbf{U}'_K \mathbf{D}^{-1} (\mathbf{r}_t - \boldsymbol{\mu}). \qquad [2.11]$$

The estimated eigenfactors are zero-mean with a diagonal covariance matrix $\mathbf{\Gamma}_K$. Each T-dimensional vector $\{\widehat{\mathbf{f}}_k\}_{k=1}^N$ of $\widehat{\mathbf{F}}$ is a linear combination of the asset returns and can be interpreted as diversified and orthogonal portfolios in which we must invest so as to capture commonalities across the assets.

One major difficulty is therefore to estimate the value of K. Some of the most commonly used techniques coming from the information theory are described in the following section.

2.4. Model order choice

Identifying K, the number of common factors, is one of the key tasks of factor models. After performing an eigen-decomposition of the covariance matrix estimate, the simplest procedure is to select the K largest eigenvalues given that various tests exist for determining if an eigenvalue is large:

1) Looking at the percentage of the total power contained in λ_k:

$$\frac{\lambda_k}{\mathrm{tr}(\hat{\Sigma})} > \eta_1. \qquad [2.12]$$

2) Computing the ratio of the $(k+1)$th eigenvalue to the kth eigenvalue which has been declared to be large:

$$\frac{\lambda_{k+1}}{\lambda_k} > \eta_2. \qquad [2.13]$$

It is obvious that the result of these tests will depend on the values chosen for η_1 or η_2. More general criteria might be useful and two families of criteria appear: (1) the traditional and (2) those defined on large panel data and more specific to the factor model identification.

2.4.1. *Information criteria*

The first category of tests to detect the value of K are referred to as sequential hypothesis (SH) tests or sphericity tests, originating from the statistical field (see [AND 63]). The objective is to find the likelihood ratio between the hypothesis that the $N - K$ smallest eigenvalues are equal versus the hypothesis that only the $N - K - 1$ smallest eigenvalues are equal. Anderson [AND 63] showed that for

$T \gg N$, the $N - K$ smallest eigenvalues clustered all around a small value s^2 and showed that a sufficient statistic is:

$$L(d) = T\,(N - d)\,\ln\left(\frac{\frac{1}{N-d}\sum_{i=d+1}^{N}\lambda_i}{\left(\prod_{i=d+1}^{N}\lambda_i\right)^{\frac{1}{N-d}}}\right). \qquad [2.14]$$

This ratio uses the ratio of the arithmetic mean of the $N - d$ smallest eigenvalues to the geometric mean of the $N - d$ smallest eigenvalues. If the $N - d$ smallest eigenvalues are equal, then $L(d) = 0$. Asymptotically, statistics [2.14] correspond to a $\chi^2((N - d)^2 - 1)$-distributed variable, so that given a confidence interval and sequential tests a value for d can be determined.

A second category of tests adds a penalty function to $L(d)$ which takes into account the degree of freedom of the model. It results in a function of d to be minimized and the classical Akaike information criteria (AIC) and minimum description length (MDL) tests can be written in such a way, that is $L(d) + p(d)$ [TRE 02, RIS 78].

2.4.1.1. *Akaike information criteria*

Akaike [AKA 74] introduced an information-theoretic criterion which is well known as the AIC. This test can be conveniently written as:

$$AIC(d) = L(d) + d\,(2N - d), \qquad [2.15]$$

so that we get the AIC estimated value $\hat{d}_{\text{aic}}$ by minimizing [2.15]:

$$\hat{d}_{\text{aic}} = \underset{d}{\operatorname{argmin}}\, AIC(d). \qquad [2.16]$$

It has been shown that AIC is inconsistent and, asymptotically, tends to overestimate the number of sources. However, for small T, the AIC generally has a higher probability of a correct decision if compared with the results obtained with the MDL test.

2.4.1.2. *Minimum description length*

MDL differs from AIC by the value of the penalty function added to the Anderson's sufficient statistic $L(d)$:

$$MDL(d) = L(d) + \frac{\ln(T)}{2} \left[d\,(2N - d) + 1 \right], \qquad [2.17]$$

so that we get the MDL estimated value $\hat{d}_{\mathrm{mdl}}$ by minimizing [2.17]:

$$\hat{d}_{\mathrm{mdl}} = \underset{d}{\operatorname{argmin}}\, MDL(d). \qquad [2.18]$$

As opposed to the AIC, MDL estimate is consistent with the time dimension, that is $T \to \infty$, MDL estimate approaches K. In practice, MDL is therefore preferred to AIC and Bayes information criterion (BIC).

2.5. Appendix 1: Covariance matrix estimation

Subspace methods require the computation of the covariance matrix of the observations which is mostly estimated by the SCM. SCM is the solution of the maximum likelihood approach under the multivariate Gaussian assumption but suffers from non-robustness in the case of outliers in returns. In this section, we first give the expressions and the properties for the sample mean and SCM and then present a family of robust estimators, the M-estimators.

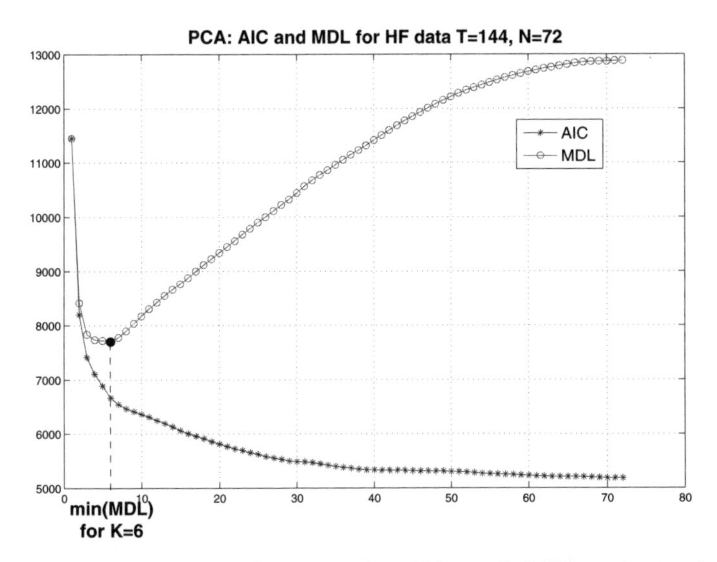

Figure 2.3. *Comparison between the AIC and MDL criteria in the determination of the number of factors in a set of $N = 72$ Hedge Funds monthly returns of size $T = 144$. MDL indicates more clearly the cutoff point in the data set ($K = 6$ factors) whereas AIC is a monotonic decreasing function*

2.5.1. *Sample mean*

The sample mean gives an unbiased and consistent estimation of the mean of a vector and is computed as the arithmetic mean of its values:

$$\widehat{\mu} = T^{-1}\,\mathbf{R}'\,\mathbf{1}_T = \begin{bmatrix} T^{-1}\sum_{t=1}^{T} r_{1,t} \\ T^{-1}\sum_{t=1}^{T} r_{2,t} \\ \vdots \\ T^{-1}\sum_{t=1}^{T} r_{N,t} \end{bmatrix}, \qquad [2.19]$$

where $\mathbf{1}_T$ denotes a T-dimensional vector of 1.

This estimator is:

– unbiased: $\mathbb{E}(\widehat{\mu}_j) = T^{-1} \sum_{t=1}^{T} \mathbb{E}(r_{j,t}) = \mu_j$,

– consistent: $\widehat{\mu} \xrightarrow{\text{w. prob. 1}} \mu$.

It coincides with the maximum likelihood estimator if the vector of observations is a Gaussian white random vector.

2.5.2. *Sample covariance matrix*

The maximum likelihood estimate $\widehat{\boldsymbol{\Sigma}}$ of $\boldsymbol{\Sigma}$ in the Gaussian case is:

$$\widehat{\boldsymbol{\Sigma}} = T^{-1} (\mathbf{R} - \mathbf{1}_N \widehat{\boldsymbol{\mu}}')' (\mathbf{R} - \mathbf{1}_N \widehat{\boldsymbol{\mu}}'). \quad [2.20]$$

This estimator is:

– consistent: $\widehat{\boldsymbol{\Sigma}} \xrightarrow{\text{w. prob. 1}} \boldsymbol{\Sigma}$,

– biased: $\mathbb{E}\left(\frac{T}{T-1} \widehat{\boldsymbol{\Sigma}}\right) = \boldsymbol{\Sigma}$.

So, if we define $\mathbf{S} = \frac{T}{T-1} \widehat{\boldsymbol{\Sigma}}$, then $\mathbf{S}$ is the so-called unbiased SCM and we have:

$$\mathbf{S} = \frac{1}{T-1} \sum_{t=1}^{T} (\mathbf{r}_t - \widehat{\boldsymbol{\mu}}) (\mathbf{r}_t - \widehat{\boldsymbol{\mu}})', \quad [2.21]$$

$$\mathbf{S} = \frac{1}{T-1} \mathbf{R}' \mathbf{Q} \mathbf{R} = \frac{1}{T-1} \mathbf{R}' (\mathbf{I} - T^{-1} \mathbf{1}_T \mathbf{1}_T') \mathbf{R}, \quad [2.22]$$

$$\mathbf{R}' \mathbf{R} = (T-1) \mathbf{S} + T \widehat{\boldsymbol{\mu}} \widehat{\boldsymbol{\mu}}'. \quad [2.23]$$

COMMENT 2.1.– If the observations are zero-mean or if the mean vector μ is known, then the maximum likelihood estimate $\widehat{\boldsymbol{\Sigma}}$ of the covariance matrix is unbiased and $\mathbf{S} = \widehat{\boldsymbol{\Sigma}}$.

Some asymptotic results:

$$\sqrt{T}\,(\widehat{\boldsymbol{\mu}} - \boldsymbol{\mu}) \xrightarrow{d} \mathbf{z} \sim \mathcal{N}_N(\mathbf{0}, \boldsymbol{\Sigma}), \quad [2.24]$$

$$T\,(\widehat{\boldsymbol{\mu}} - \boldsymbol{\mu})'\,\boldsymbol{\Sigma}^{-1}\,(\widehat{\boldsymbol{\mu}} - \boldsymbol{\mu}) \xrightarrow{d} \chi_r^2, \quad [2.25]$$

$$T\,(\widehat{\boldsymbol{\mu}} - \boldsymbol{\mu})'\,\mathbf{S}^{-1}\,(\widehat{\boldsymbol{\mu}} - \boldsymbol{\mu}) \xrightarrow{d} \mathcal{T}_{N,T-1}^2. \quad [2.26]$$

In [2.25], r denotes the rank of $\boldsymbol{\Sigma}$ (if matrix is full rank, then $r = N$). $\mathcal{T}_{N,T-1}^2$ is the Hotelling's T-squared statistic with parameters N and $T - 1$ [HOT 31]. Hotelling's T-squared statistic, developed by Harold Hotelling, is a generalization of Student's T-test statistic and is often used in multivariate hypothesis testing. When a random variable $X \sim \mathcal{T}_{p,n}^2$, then $\dfrac{n-p+1}{np} X \sim F_{p,n-p+1}$ where $F_{p,n-p+1}$ is the F-distribution with parameters p and $n - p + 1$ also called Snedecor's F-distribution or Fisher–Snedecor distribution (see appendices A1.4 and A1.5).

Gaussian case: If the observations are Gaussian multivariate $\mathcal{N}(\boldsymbol{\mu}, \boldsymbol{\Sigma})$, then the maximum likelihood estimates $\widehat{\boldsymbol{\mu}}$ and $\widehat{\boldsymbol{\Sigma}}$ are derived by minimizing:

$$\ln|\boldsymbol{\Sigma}| + \operatorname{tr}\left(\frac{1}{T}\mathbf{V}\,\boldsymbol{\Sigma}^{-1}\right) + (\widehat{\boldsymbol{\mu}} - \boldsymbol{\mu})'\,\boldsymbol{\Sigma}^{-1}\,(\widehat{\boldsymbol{\mu}} - \boldsymbol{\mu}), \quad [2.27]$$

when $(T - 1) \geq N$ and where $\mathbf{V} = (T - 1)\,\mathbf{S}$. Since the last term is positive, we only have to minimize:

$$\ln|T\,\mathbf{V}^{-1}\,\boldsymbol{\Sigma}| + \operatorname{tr}\left(\frac{1}{T}\mathbf{V}\,\boldsymbol{\Sigma}^{-1}\right). \quad [2.28]$$

Then, we have:

$$\widehat{\boldsymbol{\mu}} \sim \mathcal{N}(\boldsymbol{\mu}, \boldsymbol{\Sigma}/T), \qquad [2.29]$$

$$(T-1)\,\mathbf{S} \sim \mathcal{W}(T-1, \boldsymbol{\Sigma}), \qquad [2.30]$$

$$\widehat{\boldsymbol{\mu}} \perp\!\!\!\perp \boldsymbol{\Sigma}. \qquad [2.31]$$

where $\mathcal{W}(n, \mathbf{A})$ denotes the Wishart distribution with parameters n and $\mathbf{A}$ (see appendix A1.6 or [BIL 99] for more details).

When the Gaussian assumption is not verified, then the SCM estimate may be corrupted by outliers and deviate from optimality so that methods using a robust approach for estimation should be preferred. Robustness of an estimator may be quantified by stable estimation results when the observed data deviate more or less slightly from the assumed model within a relatively large confidence interval. The literature in robust methods for estimation is large and was probably first stated in the 1960's by John Wilder Tukey [TUK 60, TUK 62]. Peter J. Huber in [HUB 64] and [HUB 67] considered that given a distribution model, the true distribution lies in a neighborhood of the model, which may describe a theoretical approach for robustness. The M-estimators are therefore introduced as a generalization of the maximum likelihood approach, which considers that the assumed model may be erroneous. In the same manner, Huber's *gross error model* assumes that within a perfectly known distribution G, a piece of datum, say $0 \leq \epsilon \leq 1$, comes from an unknown distribution H, so that the complete characterization of the datum is given by distribution $F = (1-\epsilon)\,G + \epsilon\,H$.

In the following section, we discuss the family of the so-called M-estimators [HUB 64, HUB 77, MAR 76], which offers an alternative to the classical SCM and comes also from the maximum likelihood theory [CON 02, GIN 02].

2.5.3. *Robust covariance matrix estimation: M-estimators*

Assuming that the N columns $\{\mathbf{r}\}_{j=1}^{N}$ of $\mathbf{R}$ are independent, the M-estimator of $\boldsymbol{\Sigma}$ is defined as the solution of the following equation:

$$\widehat{\mathbf{M}} = \frac{1}{T} \sum_{t=1}^{\prime} u\left(\mathbf{r}_t' \widehat{\mathbf{M}}^{-1} \mathbf{r}_t\right) \mathbf{r}_t \mathbf{r}_t', \qquad [2.32]$$

where the function $u(.)$ may follow the following conditions [MAR 76]:

1) $u(s)$ is non-negative, non-increasing and continuous for $s \geq 0$;

2) Defining $\psi(s) = s\,u(s)$, $\forall s \geq 0$, $\psi(.)$ is bounded with $K = \sup_{s\geq 0} \psi(s)$;

3) ψ is non-decreasing and is strictly increasing in the interval where $\psi < K$;

4) There exists s_0 such that $\phi(s_0^2) > N$.

2.5.3.1. *Huber's M-estimator*

Huber's *M*-estimator is defined for the specific weighting function u:

$$u(s) = \frac{1}{\beta} \min_{s \in \mathbb{R}^+} \left(1, \frac{a}{s}\right) = \frac{1}{\beta} \left(\mathbb{1}_{s \leq a} + \frac{a}{s} \mathbb{1}_{s > a}\right), \qquad [2.33]$$

where $\mathbb{1}(x)$ is the indicator function defined by $\mathbb{1}(x) = 1$ if x is verified, 0 otherwise.

Huber's estimator is then a solution to:

$$\widehat{\mathbf{M}}_{huber} = \frac{1}{T\beta} \sum_{t=1}^{\prime} \left[\mathbf{r}_t \mathbf{r}_t' \mathbb{1}_{\mathbf{r}_t' \widehat{\mathbf{M}}_{huber}^{-1} \mathbf{r}_t \leq a} + a \frac{\mathbf{r}_t \mathbf{r}_t'}{\mathbf{r}_t' \widehat{\mathbf{M}}_{huber}^{-1} \mathbf{r}_t} \mathbb{1}_{\mathbf{r}_t' \widehat{\mathbf{M}}_{huber}^{-1} \mathbf{r}_t > a} \right]. \qquad [2.34]$$

Function [2.33] is shown in Figure 2.4. Parameters a and β are adjustable: a allows us to control the percentage of data to be attenuated and β allows us to control the distance between the asymptotic value of the estimator and its theoretical value. If a is large, then Huber's estimator behaves like the SCM: it means that the amount of data to be considered as outliers or outside the main distribution is negligible, so that all the samples are taken into account in the estimation. On the other hand, if a is small, then Huber's estimator tends toward the fixed point (FP) estimator. The latter is described hereafter.

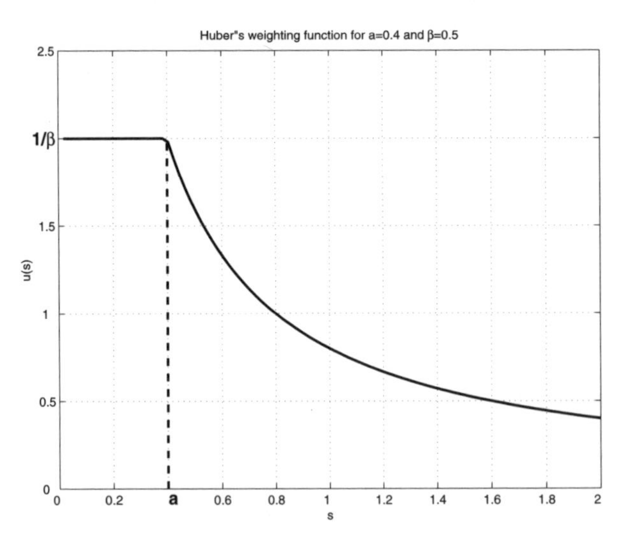

Figure 2.4. *Weighting function u of the Huber's M-estimator when $a = 0.4$ and $\beta = 0.5$*

2.5.3.2. *The fixed-point (FP) estimator*

The so-called FP estimate $\widehat{\mathbf{M}}_{fp}$ of $\boldsymbol{\Sigma}$ is defined as a FP of a function $f_{N,\boldsymbol{\Sigma}}$ such that:

$$f_{N,\boldsymbol{\Sigma}} \begin{cases} \mathcal{D} \longrightarrow \mathcal{D} \\ \mathbf{M} \longrightarrow \dfrac{N}{T} \displaystyle\sum_{t=1}^{\prime} \dfrac{\mathbf{r}_t \mathbf{r}_t'}{\mathbf{r}_t' \mathbf{M}^{-1} \mathbf{r}_t} \end{cases}, \qquad [2.35]$$

where $\mathcal{D} = \{\mathbf{M} \in \mathcal{M}_N(\mathbb{C}) | \mathbf{M}^H = \mathbf{M}, \mathbf{M}$ positive definite$\}$ with $\mathcal{M}_N(\mathbb{C}) = \{N \times N$ matrices with elements in $\mathbb{C}\}$.

Equation [2.35] or equation $\widehat{\boldsymbol{\Sigma}}_{fp} = f_{N, \widehat{\boldsymbol{\Sigma}}_{fp}}$ has a solution of the form $\alpha \boldsymbol{\Sigma}$ where α is an arbitrary scaling factor (see [PAS 08] for a detailed explanation). The name of this estimator comes from the eponymous algorithm that is used for its computation [PAS 08, TYL 83]. It has been shown that this estimator is consistent and unbiased [PAS 06]. Under particular statistical assumption for vector r, it may be possible to get its asymptotic distribution (e.g. if r is a complex Gaussian compound vector (like a spherically invariant random vector (SIRV)), then [2.35] tends, in law, toward a complex Gaussian vector [PAS 05]).

As mentioned above, the FP solution of equation [2.35] is up to a scalar α. To get one of the solutions, we may impose a specific normalization, such that $\mathrm{tr}(\widehat{\boldsymbol{\Sigma}}_{fp}) = N$. Moreover, this norm may also be imposed on the theoretical covariance matrix.

The performance of the SCM, Huber's and FP estimators is compared in Figure 2.5. Data are simulated from a standard Gaussian distribution with $T = 250$ observations and $N = 100$ assets. Some outliers are randomly introduced in the data and come from a Bernoulli–Gauss process of variance 15. We have assumed that 50% of each of the N observation vectors is corrupted with outliers. Moreover, in order to compute Huber's M-estimator, we have assumed that only 60% of the data are not corrupted, so that the parameter α is the value of a χ^2 variable with N degree of freedom within a 60% confidence interval.

Figure 2.5 shows that the SCM estimate is sensitive to outliers whereas the FP estimate is fully robust to outliers. Huber's M-estimate is somewhere between the both, depending on parameters a and β. As a reference, we have added the SCM estimate on the uncorrupted data.

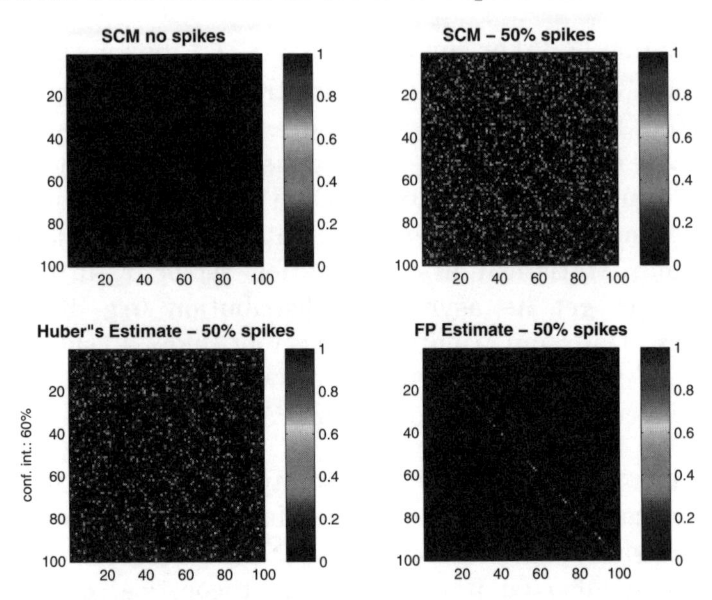

Figure 2.5. *Comparison between the SCM, Huber's and FP estimators of the covariance matrix of $T = 250$ i.i.d. normal returns of size $N = 100$ sampled from a $\mathcal{N}(\mathbf{0}, \mathbf{I})$. Some outliers (50% of the data) are randomly introduced in the observations sampled from a Bernoulli–Gauss process of variance 15. We have assumed, for Huber's estimator, that 60% of the data come from a Gaussian distribution (see color plate section)*

2.6. Appendix 2: Similarity of the eigenfactor selection with the MUSIC algorithm

As mentioned in section 2.3.2, the MUSIC algorithm [STO 89] uses the same methodology as principal component analysis does, that is it is based on the eigen-decomposition of the covariance matrix of the observations. Nevertheless, MUSIC algorithm is used for data described by the following model:

$$\mathbf{y}_t = \mathbf{A}(\theta)\, \mathbf{x}_t + \mathbf{e}_t, \qquad [2.36]$$

where $\mathbf{A}(\boldsymbol{\theta})$ is an $m \times n$ Vandermonde matrix, $\mathbf{x}_t$ is an n-dimensional vector representing the amplitude of the observed signal $\mathbf{y}_t$ and $\mathbf{e}_t$ is a zero-mean m-dimensional vector of white Gaussian noise. In this model, the unknown parameters are the $n < m$ arrival angles present in $\boldsymbol{\theta}$ and the problem is therefore: (1) to estimate the number of sources n, which is often supposed fixed, (2) to estimate the amplitudes of the signal: when the vector $\boldsymbol{\theta}$ is estimated, then the amplitudes are estimated by using a maximum likelihood estimation (MLE) approach and (3) to estimate $\boldsymbol{\theta}$ that uses the eigen-decomposition of the covariance matrix.

The covariance matrix of $\mathbf{y}$ is therefore:

$$\mathbf{M}_y = \mathbb{E}(\mathbf{y}\,\mathbf{y}^H) = \mathbf{A}(\boldsymbol{\theta})\,\mathbf{M}_x\,\mathbf{A}(\boldsymbol{\theta})^H + \sigma^2\,\mathbf{I}, \qquad [2.37]$$

where the superscript H is the transpose and conjugate operation, $\mathbf{M}_x = \mathbb{E}(\mathbf{x}\,\mathbf{x}^H)$ is the $n \times n$ signal auto-correlation matrix and σ^2 is the noise variance. Assuming that $n < m$ signals are present, then the n eigenvectors associated with the n largest eigenvalues define the signal subspace (the column space of $\mathbf{A}$), and the $m - n$ other eigenvalues define the noise subspace $\mathcal{U}_v$. As the signal subspace is orthogonal to the noise subspace, the strongest frequencies correspond to the peaks of the following function:

$$J(w) = \frac{1}{\mathbf{a}(w)^H\,\mathcal{U}_v\,\mathcal{U}_v^H\,\mathbf{a}(w)}. \qquad [2.38]$$

MUSIC algorithm assumes that the number of samples m and the number of sources are known and its efficiency is measured by the ratio between the smallest variance (given by the Cramér–Rao bound – CRLB) [CRA 46] and the variance of the MUSIC estimator, that is $eff = Var_{CRLB}/Var_{music}$.

2.7. Appendix 3: Large panel data

In the asymptotic case where the number of observations T and the number of assets N are large, Bai and Ng [BAI 02] have developed an econometric theory to determine the number of latent factors. They consider that the data have an r factor representation:

$$\boldsymbol{X} = \boldsymbol{F}\boldsymbol{\Gamma}' + \boldsymbol{e}, \qquad [2.39]$$

where $\boldsymbol{X}$ is the $T \times N$ matrix of observed data, $\boldsymbol{F}$ is the $T \times r$ matrix of factors, $\boldsymbol{\Gamma}$ is the $N \times r$ matrix of factor loadings and e is the $T \times N$ matrix of idiosyncratic noise. It may be rewritten for each asset $i = 1, \cdots, N$ and each date $t = 1, \cdots, T$ as follows:

$$X_{i,t} = \boldsymbol{\lambda}_i' \boldsymbol{F_t} + e_{i,t}. \qquad [2.40]$$

In this representation, it is supposed that the factors, their loadings and the idiosyncratic errors are not observable. Their approach differs from the theory developed for classical factor models, which does not apply when both N and $T \to \infty$. Inference on r can be theoretically based on the eigenvalues of the covariance matrix estimate since within the r factor representation [2.40] the first r largest population eigenvalues diverge as N increases to infinity, but the $(r + 1)th$ eigenvalue is bounded [CHA 83]. And it can be shown that all non-zero sample eigenvalues increase with N, a test based on the sample eigenvalues is thus not feasible. A likelihood ratio test can also be used to select the number of factors if the residual noise e_t is assumed to be Gaussian distributed.

The authors estimate the common factors by the method of asymptotic principal components. The true number of factor r is unknown but fixed. They start with an arbitrary number

k ($k < \min(N, T)$) and then solve the following optimization problem:

$$V(k) = \min_{\Gamma, F^k} \frac{1}{NT} \sum_{i=1}^{N} \sum_{t=1}^{T} (X_{it} - \lambda_i^k F_t^k)^2, \qquad [2.41]$$

where the superscript k indicates the current order of the model in the iterated process.

The optimization problem [2.41] is subject to the normalization of either $\frac{\Gamma^{k'} \Gamma^k}{N} = I_k$ or $\frac{F^{k'} F^k}{T} = I_k$. Considering the latter one, then solving [2.41] is equivalent to maximizing $\mathrm{tr}(\boldsymbol{F}^{k'} (\boldsymbol{X}\boldsymbol{X}') \boldsymbol{F}^k)$. The estimated factor matrix denoted by $\widetilde{\boldsymbol{F}}^k$ is $\sqrt{T}$ times the eigenvectors corresponding to the k largest eigenvalues of the $T \times T$ matrix $\boldsymbol{X}\boldsymbol{X}'$. Given $\widetilde{\boldsymbol{F}}^k$, then $\widetilde{\boldsymbol{\Gamma}}^{k'} = \left(\widetilde{\boldsymbol{F}}^{k'} \widetilde{\boldsymbol{F}}^k\right)^{-1} \widetilde{\boldsymbol{F}}^{k'} \boldsymbol{X} = \frac{\widetilde{\boldsymbol{F}}^{k'} \boldsymbol{X}}{T}$ is the corresponding matrix of factor loadings.

2.7.1. *Large panel data criteria*

AIC, BIC and MDL are not consistent when both cross-sectional and time dimension increase. Bai and Ng [BAI 02] developed an asymptotic theory for the *approximate* factor models of Chamberlain and Rothschild [CHA 83] with large panel data. This results in consistent estimators (the so-called IC_p and PC_p estimators) that compute the number of eigenvalues larger than a threshold value, specified by a chosen penalty function. Onatski [ONA 06] modified these estimators by computing the threshold from the empirical distribution of the eigenvalues. In a similar context, Ahn and Horenstein [AHN 09] proposed the eigenvalue ratio (ER) and the growth ratio (GR) estimators, obtained simply by maximizing the ratio of two adjacent eigenvalues. Recent advances in the field of random matrix theory (RMT), which provides mathematical tools to characterize the empirical eigenvalue distribution for many symmetric matrices [ELD 05], can also be used in this context [POT 05].

2.7.1.1. *Panel C_p (PC_p and information criteria (IC_p)) estimators [BAI 02]*

Let us first recall equation [2.41] that denotes the sum of squared residuals when k factors are estimated:

$$V(k, \hat{F}^k) = \min_{\Gamma} \frac{1}{NT} \sum_{i=1}^{N} \sum_{t=1}^{\prime} (X_{it} - \lambda_i^{k'} \hat{F}_t^k)^2.$$

Estimating the number of factors is related to finding a penalty function $g(N,T)$ such that criteria of the form:

$$PC(k) = V(k, \hat{\boldsymbol{F}}^k) + k\, g(N,T), \qquad [2.42]$$

can consistently estimate K with a bounded integer K_{max} such that $K < K_{max}$. In the same way, the authors show that the class of criteria defined by:

$$IC(k) = \ln(V(k, \hat{\boldsymbol{F}}^k)) + k\, g(N,T), \qquad [2.43]$$

can also consistently estimate K. Various forms of PC_p and IC_p are then obtained from [2.42] and [2.43] and give:

$$PC_{p_1}(k) = V(k, \hat{\boldsymbol{F}}^k) + k\, \hat{\sigma}^2 \left(\frac{N+T}{NT}\right) \ln\left(\frac{NT}{N+T}\right), \qquad [2.44]$$

$$PC_{p_2}(k) = V(k, \hat{\boldsymbol{F}}^k) + k\, \hat{\sigma}^2 \left(\frac{N+T}{NT}\right) \ln C_{NT}^2, \qquad [2.45]$$

$$PC_{p_3}(k) = V(k, \hat{\boldsymbol{F}}^k) + k\, \hat{\sigma}^2 \left(\frac{\ln C_{NT}^2}{C_{NT}^2}\right), \qquad [2.46]$$

$$IC_{p_1}(k) = \ln(V(k, \hat{\boldsymbol{F}}^k)) + k \left(\frac{N+T}{NT}\right) \ln\left(\frac{NT}{N+T}\right), \qquad [2.47]$$

$$IC_{p2}(k) = \ln(V(k, \hat{\boldsymbol{F}}^k)) + k\left(\frac{N+T}{NT}\right)\ln C_{NT}^2, \quad [2.48]$$

$$IC_{p3}(k) = \ln(V(k, \hat{\boldsymbol{F}}^k)) + k\left(\frac{\ln C_{NT}^2}{C_{NT}^2}\right), \quad [2.49]$$

where $\hat{\sigma}^2$ is set to be a consistent estimate of the variance of the residuals and $C_{NT} = \min(\sqrt{N}, \sqrt{T})$. The class of PC_p criteria generalizes the C_p criterion of Mallows [MAL 73] developed in a different context. Note also that the penalty functions $g(N,T)$ in equations [2.44]–[2.49] depend on both N and T, which is not the case for the AIC and BIC. Moreover, these criteria show that the penalty functions satisfy the two following conditions, necessary to get consistent estimate of K : (1) $g(N,T) \to 0$, and (2) $C_{NT}^2\, g(N,T) \to \infty$ as $N, T \to \infty$ (see [BAI 02] for the details).

Let us also note that the classical AIC and BIC criteria can also be written as follows:

$$AIC(k) = V(k, \hat{\boldsymbol{F}}^k) + k\,\hat{\sigma}^2\left(\frac{2}{T}\right), \quad [2.50]$$

$$BIC(k) = V(k, \hat{\boldsymbol{F}}^k) + k\,\hat{\sigma}^2\left(\frac{\ln T}{T}\right). \quad [2.51]$$

2.7.1.2. *Eigenvalue distribution test: random matrix theory*

Statistical theory usually presumes a constant number of dimensions or at least $T/N \to \infty$. The quantity $q = T/N$ can be interpreted as average sample size per dimension or as effective sample size. Unfortunately, large sample properties fail if q is small even if T is large. RMT is a branch of statistical physics that deals with this case of high-dimensional data. RMT is mainly concerned with the distribution of eigenvalues of randomly generated matrices. An important result is that under the hypothesis of independent and identically distributed matrix elements, the distribution of the eigenvalues converges to a specified law that does not depend on the distribution of the matrix elements but primarily on $q = T/N$.

Since the SCM [2.21] is a random matrix (see equations [2.26] and [2.30]), the results of RMT can be applied in the case of normally distributed data. In cases where the data are generalized elliptically distributed, the results of RMT are no longer applicable if we use the SCM, but may find some evidence by using the spectral estimator instead. Results on generalized elliptical distributions and the spectral estimator of the covariance matrix can be found in [TYL 83], [TYL 87a] and [TYL 87b] where it is shown that for such distributions the spectral estimator corresponds to an M-estimator.

This section focuses on the basis of the RMT. The reader may refer to [FRA 04], [POT 05], [HAR 07] and [BAI 10] for a more complete and advanced description of the RMT.

Suppose that we use the SCM computed on the standardized matrix of returns $\tilde{\mathbf{R}}$ as described in equation [2.4]. The resulting estimated covariance matrix $\mathbf{S}$ has therefore random elements with an assumed variance σ^2. Then, in the limit $T, N \to \infty$, keeping the ratio $q = T/N \geq 1$ constant, the density of the eigenvalues of $\mathbf{S}$ is given by:

$$\rho_q(\lambda) = \frac{q}{2\pi\sigma^2} \frac{\sqrt{(\lambda_+ - \lambda)^+ (\lambda - \lambda_-)^+}}{\lambda}, \qquad [2.52]$$

where the maximum and the minimum eigenvalues are given by:

$$\lambda_\pm = \sigma^2 \left(1 \pm \sqrt{\frac{1}{q}}\right)^2, \qquad [2.53]$$

and $(x - a)^{+} = max(0, x - a)$. The density $\rho(\lambda)$ is known as the Marčenko–Pastur density [MAR 67, PAS 11]. Figure 2.6 shows the form of the Marčenko–Pastur density for different values of $q = T/N$.

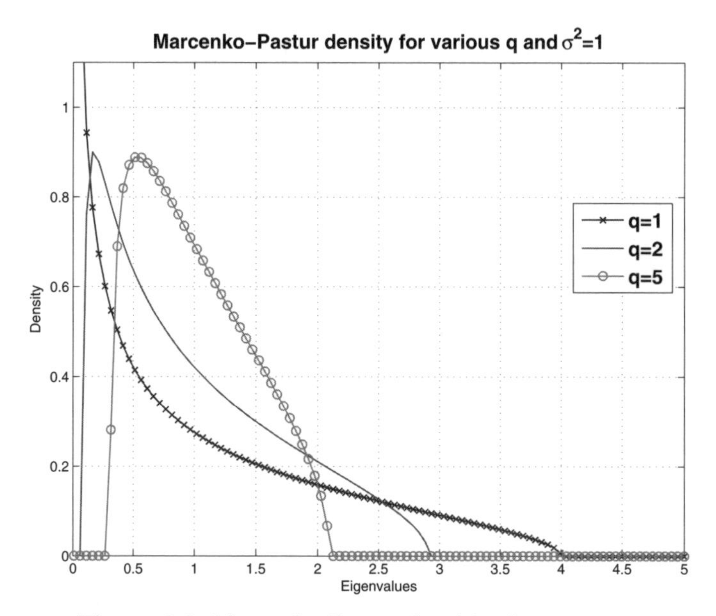

Figure 2.6. *Marčenko–Pastur densities for* $q = 1, 2, 5$

As is illustrated in Figures 2.7–2.9 even for rather small matrices, the theoretical limiting density approximates the actual density very well. In these three figures, the ratio $q = T/N$ is successively equal to 5, 2 and 1.

As mentioned in various studies, the Marčenko–Pastur density is a very good approximation of the density of eigenvalues of the correlation matrix of randomized vectors even though the density is fat-tailed. The question is therefore how to select the significant eigenvalues and then to determine the order of the model or the number of factors and to interpret the corresponding eigenvectors. The latter point is related to the observed data and it has been pointed out, for example, that the eigenvector corresponding to the largest eigenvalue is representative of the market factor as it covers the major part of the variance of the returns. The other eigenvectors may be compared with observed time series in the market as it was done in [DAR 11].

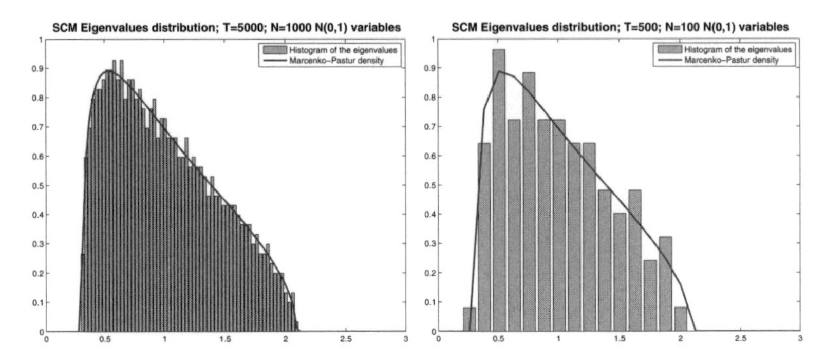

Figure 2.7. *Distribution of the eigenvalues of the SCM for $q = 5$ computed from i.i.i. random normal returns using ($T = 5{,}000, N = 1{,}000$) and ($T = 500, N = 100$). The corresponding Marčenko–Pastur density is superimposed*

In a more general setting, selection of the significant eigenvalues is done through a thresholding procedure: all eigenvalues above some number $\lambda*$ are considered informative, otherwise eigenvalues relate to noise. To reconstruct a filtered covariance matrix estimate or to construct the corresponding eigenfactors, all noise-related eigenvalues may be replaced with a constant. This yields to renormalize the eigenvalues by the total sum. Then, the denoised covariance matrix estimate is obtained by the inverse procedure of its diagonalization.

Let us recall that each eigenvalue relates to the variance of a portfolio of assets. A very small eigenvalue means that there exists a portfolio of assets with very small out-of-sample variance, which is probably not realistic.

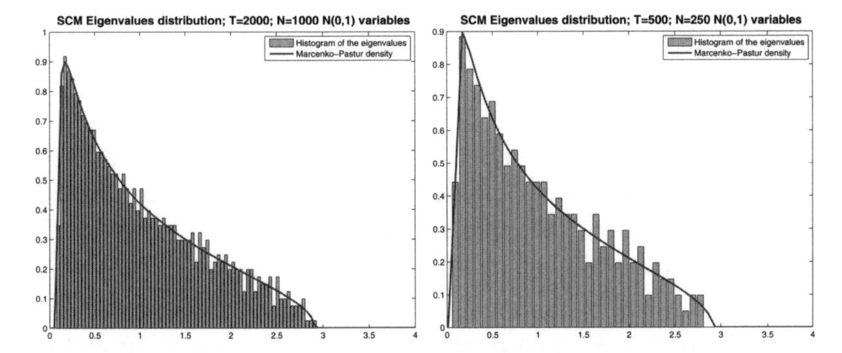

Figure 2.8. *Distribution of the eigenvalues of the SCM for $q = 2$ computed from i.i.i. random normal returns using ($T = 2{,}000, N = 1{,}000$) and ($T = 500, N = 250$). The corresponding Marčenko–Pastur density is superimposed*

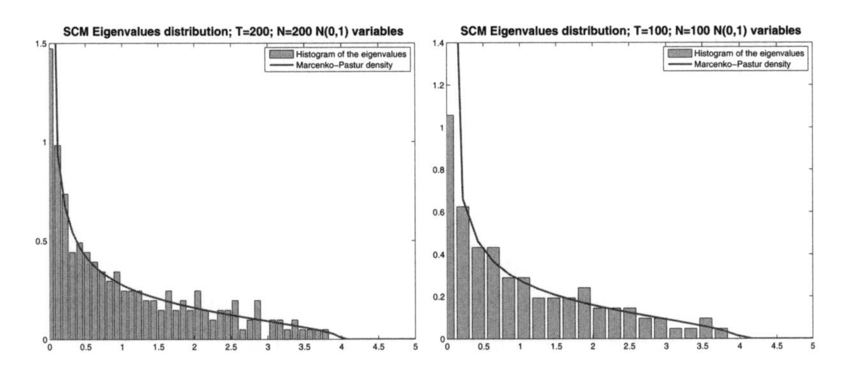

Figure 2.9. *Distribution of the eigenvalues of the SCM for $q = 1$ computed from i.i.i. random normal returns using ($T = 500, N = 500$) and ($T = 200, N = 200$). The corresponding Marčenko–Pastur density is superimposed*

2.8. Chapter 2 highlights

Best factor models

– The three-factor model of Fama and French [FAM 93] includes two additional factors to Sharpe's market model: small minus big (SMB) market cap and high minus low (HML) book-to-market value.

– The four-factor model of Carhart [CAR 97] is an extension of the Fama and French model including a momentum factor.

– The Chen *et al.* model [CHE 86] is a model that includes several macro-economic factors that explain the returns of stocks in the US market.

– The eight-factor model of Fung and Hsieh [FUN 01], designed to explain trend-followers' hedge funds, includes eight risk-based factors: three specific portfolios invested in commodities, bonds and foreign exchange, two equity-oriented risk factors, two bond-oriented factors and an equity index of emerging stocks.

Sample estimates

Given the $T \times N$ matrix of observations $\mathbf{R}$ with mean μ and covariance matrix Σ, the consistent unbiased sample estimates for μ and Σ are, respectively, given by:

$$T\,\hat{\boldsymbol{\mu}} = \mathbf{R}'\,\mathbf{1}, \text{and}$$

$$(T-1)\,\mathbf{S} = \mathbf{R}'\,\mathbf{R} - T\,\hat{\boldsymbol{\mu}}\,\hat{\boldsymbol{\mu}}.$$

Wishart distribution

The Wishart distribution $\mathcal{W}(K, \mathbf{A})$ is the distribution of $\sum_{k=1}^{K} \mathbf{z}_k\, \mathbf{z}_k'$ where the $\mathbf{z}_k$s are independent, identically distributed (i.i.d.) according to a Gaussian distribution with zero-mean and covariance **A**.

***M*-estimates**

$$\hat{\mathbf{M}} = \frac{1}{T}\sum_{t=1}^{T} u\left(\mathbf{r}_t' \hat{\mathbf{M}}^{-1} \mathbf{r}_t\right) \mathbf{r_t} \mathbf{r_t}'$$

Huber's M-estimator

$$u(s) = \frac{1}{\beta} \min_{s \in \mathbb{R}^+} \left(1, \frac{a}{s}\right) = \frac{1}{\beta}\left(\mathbb{1}_{s \leq a} + \frac{a}{s}\mathbb{1}_{s>a}\right)$$

Fixed-point estimate

$$f_{N,\boldsymbol{\Sigma}} = \frac{N}{T}\sum_{t=1}^{T} \frac{\mathbf{r}_t \mathbf{r}_t'}{\mathbf{r}_t' \mathbf{M}^{-1} \mathbf{r}_t}$$

Eigen-decomposition

$$\hat{\boldsymbol{\Sigma}} = \sum_{k=1}^{N} \lambda_i \, \mathbf{u}_i \, \mathbf{u}_i' = \mathbf{U} \, \boldsymbol{\Gamma} \, \mathbf{U}'$$

K eigenfactors

$$\hat{\mathbf{f}}_t = \mathbf{U}_K' \, \mathbf{D}^{-1} \left(\mathbf{r}_t - \boldsymbol{\mu}\right)$$

Anderson's sufficient statistic

$$L(d) = T\,(N-d)\,\ln\left(\frac{\frac{1}{N-d}\sum_{i=d+1}^{N} \lambda_i}{\left(\prod_{i=d+1}^{N} \lambda_i\right)^{\frac{1}{N-d}}}\right)$$

Akaike information criteria

$$AIC(d) = L(d) + d\,(2\,N - d)$$

$$\hat{d}_{aic} = \underset{d}{\operatorname{argmin}}\, AIC(d)$$

Minimum description length

$$MDL(d) = L(d) + \frac{\ln(T)}{2}\,[d\,(2\,N - d) + 1]$$

$$\hat{d}_{mdl} = \underset{d}{\operatorname{argmin}}\, MDL(d)$$

Eigenvalues follow the Marčenko–Pastur density

When both $T, N \to \infty$ and given $q = T/N \geq 1$ constant, the density of the eigenvalues of S is given by the Marčenko–Pastur density:

$$\rho_q(\lambda) = \frac{q}{2\,\pi\,\sigma^2}\,\frac{\sqrt{(\lambda_+ - \lambda)^+\,(\lambda - \lambda_-)^+}}{\lambda},$$

where the maximum and the minimum eigenvalues are given by:

$$\lambda_\pm = \sigma^2\left(1 \pm \sqrt{\frac{1}{q}}\right)^2,$$

and $(x - a)^+ = max(0, x - a)$.

Chapter 3

Least Squares Estimation (LSE) and Kalman Filtering (KF) for Factor Modeling: *A Geometrical Perspective*

3.1. Introduction

Following a geometrical approach, this chapter introduces, illustrates and derives both least squares estimation (LSE) and Kalman filter (KF) estimation of the alpha and betas of a return, for a given number of factors that have already been selected. An emphasis is put on commonalities and discrepancies between least squares (LS) and KF. In particular, both techniques use as the "measurement" equation or regression the "per return" factor model that linearly links the inputs (return and factors) with the unknown parameters (alpha and betas). In a common geometrical framework, both LS and KF optimization criteria aim at minimizing the square Euclidian norm of an error-vector. While KF provides genuine recursive estimates of the alpha and betas, the LS technique cannot, since it processes the inputs in a block fashion, over a certain sliding observation window. Moreover, the window has to be long enough to ensure some numerical stability. LSE thus exhibits limited tracking capabilities. Tracking abilities of the KF are induced by an underlying recursive model (called the state equation) of the alpha and betas that is not part of LSE setup.

This chapter is organized as follows. Section 3.2 formalizes the "per return factor model" and the concept of recursive estimate of the alpha and betas. Sections 3.3–3.7 are dedicated to LSE while the contents of sections 3.8–3.12 deal with KF. A summary of the most important outcomes of the chapter is contained in section 3.15. Numerous simulation results are displayed and commented throughout of the chapter to illustrate the behaviors, performance and limitations of LSE and KF.

3.2. Why LSE and KF in factor modeling?

3.2.1. *Factor model per return*

For a single asset i, the model of return $r_{i,t}$ at time t based on the K factors $f_{t,k}$, $1 \leq k \leq K$ takes the following form [JAY 11a], $\forall 1 \leq i \leq N$:

$$r_{i,t} = \alpha_{i,t} + \sum_{k=1}^{K} b_{i,t,k}\, f_{t,k} + \epsilon_{i,t}. \qquad [3.1]$$

NOTE 3.1.– The assumptions inherent to the model [3.1] are the same as the assumptions introduced in Chapter 1. In particular, $\epsilon_{i,t}$ are zero-mean random processes with $\mathbb{E}[\epsilon_{i,t}\, \epsilon_{j,t}] = \sigma_\epsilon^2\, \delta[i-j]\, \delta[t_1 - t_2]$, $\delta[0] = 1$ and $\delta[k] = 0, k \neq 0$.

In this chapter, for the sake of simplicity, the reference to a specific asset will be implicit and the index i is thus dropped. Accordingly, we introduce $\boldsymbol{\theta}_t$ the unknown $K+1$-dimensional "parameter-vector" at time t, which collects α and the K betas (per return), such that:

$$\boldsymbol{\theta}_t' = \left[\alpha_t \; b_{t,1} \; b_{t,2} \cdots b_{t,K}\right]. \qquad [3.2]$$

If we aggregate the K factors $f_{t,k}$, $1 \leq k \leq K$ in the $(K+1)$-sized vector $\mathrm{g}_t' = \left[1 \; f_{t,1} \; f_{t,2} \cdots f_{t,K}\right]$, with its first component equal to one, then equation [3.1] can be rewritten in a more compact form called hereafter the *per return factor model* or the observation/measurement equation:

$$r_t = \mathrm{g}_t' \boldsymbol{\theta}_t + \epsilon_t. \qquad [3.3]$$

The (per asset) return r_t that is also referred to, in a more generic way, as the observation, or the measurement, or the response, is the first input of model [3.3]. The "factors" $g_{t,k}$, $0 \leq k \leq K$ (with $g_{t,0} = 1$, $g_{t,k} = f_{t,k}$ for $k \neq 0$), commonly designated as the regressors in the framework of linear regression, are the other inputs to the per asset factor model.

3.2.2. *Alpha and beta estimation per return*

In this chapter, it is assumed that the K factors $f_{t,k}$, $1 \leq k \leq K$ have already been selected. LSE and KF are two different methods, among many others, that allow the estimation of $\boldsymbol{\theta}_t$ for a given set of factors. Such an estimate is denoted $\hat{\boldsymbol{\theta}}_t$.

LSE and KF exhibit two major commonalities that are leveraged throughout the chapter. First, they share the same observation equation [3.3]. Second, their optimization criteria and subsequent solutions can be interpreted geometrically in similar ways, involving the square of specific Euclidian or "second order" distances. For both of them, the geometrical light eases the understanding of how the methods are working, how they perform and how the optimum solutions are derived.

One fundamental discrepancy differentiates LSE and KF. Thanks to "instantaneous and recursive processing" of the inputs mentioned above, KF provides a recursive estimate $\hat{\theta}_t$ that can be naturally expressed as a function of $\hat{\theta}_{t-1}$. Contrary to KF, LSE has limited "tracking capabilities" (*of dynamic investment styles like in hedge funds, for instance*), since it is based on "block processing" and does not yield any recursive estimate. As opposed to instantaneous processing, block processing requires the access to several (and not only one) measurements and factors values to come up with an estimate.

These points are detailed in sections 3.3–3.7 for block processing-based LSE and in sections 3.8–3.12 for recursive and instantaneous processing-based KF.

3.3. LSE setup

3.3.1. *Current observation window and block processing*

Operating LSE method requires us to define the "current observation" window that limits the time index of the inputs involved in the block processing needed to compute the current parameter LSE. The current observation window is denoted (t, h), where t represents the current time instant (*i.e. actually chosen as the end of the window*) and h the number of time instants comprised in this window. We have $(t, h) = \{l$, such that: $t - h + 1 \leq l \leq t\}$. We also call h the window length.

3.3.2. *LSE regression*

Since LSE leverages block processing of the responses r_l and the regressors g_l' for $l \in (t, h)$, LSE implicitly assumes that the parameter-vector of alpha and betas holds a certain value denoted $\boldsymbol{\theta}_{(t,h)}$ across the observation window (t, h). This allows the generation of the measurements r_l for $l \in (t,h)$, according to the so-called "LSE regression", derived from equation [3.3]:

$$r_l = \mathbf{g}_l' \boldsymbol{\theta}_{(t,h)} + \epsilon_l, \; t-h+1 \leq l \leq t. \qquad [3.4]$$

LSE regression [3.4] can be rewritten in a "block" fashion, as it is more appropriate to define the LSE objective criterion and derive its solution:

$$\mathbf{r}_{(t,h)} = \mathbf{G}_{(t,h)} \boldsymbol{\theta}_{(t,h)} + \boldsymbol{\epsilon}_{(t,h)}. \qquad [3.5]$$

In equation [3.5], we have $\boldsymbol{r}'_{(t,h)} = \left[r_t \; r_{t-1} \cdots r_{t-h+1}\right]$, $\mathbf{G}_{(t,h)} = rows(\mathbf{g}_l')$, $t-h+1 \leq l \leq t$ and $\boldsymbol{\epsilon}'_{(t,h)} = \left[\epsilon_t \; \epsilon_{t-1} \cdots \epsilon_{t-h+1}\right]$. The LSE "block regression" aggregates h instances of the return process r_l.

ASSUMPTION 3.1.– The window length h is assumed to be much larger than the number $K+1$ of unknown quantities, that is $h \gg K+1$ and the $[h, K+1]$ matrix $\mathbf{G}_{(t,h)}$ is assumed to be full rank (*equal to* $K+1$).

3.4. LSE objective and criterion

The LSE $\hat{\boldsymbol{\theta}}_{(t,h)}$ achieves the minimum of the squared Euclidian norm $\|.\|_{l_2}$ of the difference $\mathbf{r}_{(t,h)} = \mathbf{G}_{(t,h)} \boldsymbol{\xi}$ among all the possible real-valued $(K+1)$-dimensional vectors $\boldsymbol{\xi}$:

$$\xi^{op}_{(t,h)} = \hat{\boldsymbol{\theta}}_{(t,h)} = \arg\{ \min_{\boldsymbol{\xi} \in \mathbb{R}^{K+1}} \|\mathbf{r}_{(t,h)} - \mathbf{G}_{(t,h)} \boldsymbol{\xi}\|^2_{L^2}\}. \qquad [3.6]$$

The term $\mathbf{G}_{(t,h)} \boldsymbol{\xi}$ can be viewed as an estimate $\hat{\mathbf{r}}_{(t,h)}$ of the response-vector $\mathbf{r}_{(t,h)}$ built across the current observation window (t,h). As such, if we denote $\mathbf{c}^k_{(t,h)}$, $1 \leq k \leq K+1$, the $K+1$ column-vectors (*also called column-factors*) of the matrix $\mathbf{G}_{(t,h)}$, the term $\mathbf{G}_{(t,h)}\,\boldsymbol{\xi}$ can be rewritten as a linear combination of the $\mathbf{c}^k_{(t,h)}$, that is:

$$\mathbf{G}_{(t,h)}\,\boldsymbol{\xi} = \sum_{k=1}^{K+1} \mathbf{c}^k_{(t,h)}\,\xi_k. \qquad [3.7]$$

Consequently and as defined in [3.6], the LSE $\hat{\boldsymbol{\theta}}_{(t,h)}$ helps us to build the linear combination of the column-regressors $\mathbf{c}^k_{(t,h)}$ (factors) that best fits the measurements (*returns*) $\mathbf{r}_{(t,h)}$ in a geometrical and Euclidian sense. The geometrical interpretation of the solution is detailed in section 3.6.2. It is important to note that the LSE optimality criterion does not put any constraint on the solution $\hat{\boldsymbol{\theta}}_{(t,h)}$. In particular, the magnitude of $\hat{\boldsymbol{\theta}}_{(t,h)}$ is not prevented from becoming very large which could lead to unreliable responses estimate $\hat{\mathbf{r}}_{(t,h)}$. In such a situation, LSE has to be regularized or constrained (see Chapter 4).

3.5. How LSE is working (for LSE users and programmers)

If $\mathbf{G}_{(t,h)}$ is full rank, then the $[K+1, K+1]$ matrix $\mathbf{G}'_{(t,h)}\,\mathbf{G}_{(t,h)}$ is invertible and the LSE $\hat{\boldsymbol{\theta}}_{(t,h)}$ requires the following block processing over the observation window (t,h) (see section 3.6 for the derivation):

$$\xi^{op}_{(t,h)} = \hat{\boldsymbol{\theta}}_{(t,h)} = (\mathbf{G}'\,\mathbf{G})^{-1}_{(t,h)}\,(\mathbf{G}'\,\mathbf{r})_{(t,h)}. \qquad [3.8]$$

According to the explicit form of $\hat{\boldsymbol{\theta}}_{(t,h)}$ given in [3.8], it is clear that the tracking capabilities of the LSE are burdensome, if not limited. As a matter of fact, the computation of $\hat{\boldsymbol{\theta}}_{(t+1,h)}$ based on the next observation window $(t+1,h)$ requires going through the full block processing, that is the calculation of both $(\mathbf{G}'\,\mathbf{G})^{-1}_{(t+1,h)}$ and $(\mathbf{G}'\,\mathbf{r})_{(t+1,h)}$. These two latter entities do not exhibit any recursivity property that would allow the derivation of $\hat{\boldsymbol{\theta}}_{(t+1,h)}$ as a mere function of the current LSE $\hat{\boldsymbol{\theta}}_{(t,h)}$. This is actually a major limitation of the LSE.

3.6. Interpretation of the LSE solution

3.6.1. *Bias and variance*

Substituting the LSE regression [3.5] into the LSE solution [3.8] reveals the relationship between the estimate $\hat{\boldsymbol{\theta}}_{(t,h)}$ and the unknown $\boldsymbol{\theta}_{(t,h)}$:

$$\hat{\boldsymbol{\theta}}_{(t,h)} = \boldsymbol{\theta}_{(t,h)} + (\mathbf{G}'\,\mathbf{G})^{-1}_{(t,h)}\,(\mathbf{G}'\,\boldsymbol{\epsilon})_{(t,h)}. \qquad [3.9]$$

Equation [3.9] turns out to be the keystone of the LSE performance evaluation both in terms of bias and variance. Indeed, in keeping with [3.9], the bias $B(\hat{\boldsymbol{\theta}}_{(t,h)})$ of the LSE $\hat{\boldsymbol{\theta}}_{(t,h)}$ defined as the difference $B(\hat{\boldsymbol{\theta}}_{(t,h)}) = \boldsymbol{\theta}_{(t,h)} - \mathbb{E}(\hat{\boldsymbol{\theta}}_{(t,h)})$ is equal to zero, because the observation noise vector $\boldsymbol{\epsilon}_{(t,h)}$ is zero-mean. This means that, on average, the LSE coincides with the true parameter. But the bias reflects only the first-order moment of $\hat{\boldsymbol{\theta}}_{(t,h)}$ and subsequently displays a narrowed view of the LSE quality.

The LSE reliability also requires gauging the magnitude of its fluctuations around the true value. If $\hat{\boldsymbol{\theta}}_{(t,h)}$ can fluctuate a lot around $\boldsymbol{\theta}_{(t,h)}$, then there is a substantial probability that the LSE value $\hat{\boldsymbol{\theta}}_{(t,h)}$ falls far from $\boldsymbol{\theta}_{(t,h)}$ which makes the LSE highly unreliable, even if it is unbiased. The covariance matrix $\mathbf{S}_{(t,h)}$ of the innovation vector, $\tilde{\boldsymbol{\theta}}_{(t,h)} = \boldsymbol{\theta}_{(t,h)} - \hat{\boldsymbol{\theta}}_{(t,h)}$, helps to quantify these fluctuations. Thanks to [3.9] and the whiteness of the observation noise $\boldsymbol{\epsilon}_{(t,h)}$, with a covariance matrix equal to $\sigma^2_\epsilon\,\mathbf{I}$, we have:

$$\mathbf{S}_{(t,h)} = \mathbb{E}[(\tilde{\boldsymbol{\theta}}\,\tilde{\boldsymbol{\theta}}')_{(t,h)}] = (\mathbf{G}'\,\mathbf{G})^{-1}_{(t,h)}\,\sigma^2_\epsilon. \qquad [3.10]$$

If the parameter $\boldsymbol{\theta}_{(t,h)}$ was scalar (one dimensional), the matrix $\mathbf{S}_{(t,h)}$ would coincide with the dispersion metrics itself, that is the variance. Hereafter, we define the dispersion index of the LSE $\hat{\boldsymbol{\theta}}_{(t,h)}$ around its average $\boldsymbol{\theta}_{(t,h)}$ as the trace of $\mathbf{S}_{(t,h)}$. It is emphasized that this trace is always positive because $\mathbf{S}_{(t,h)}$ is a positive definite matrix. Denoting $\{\lambda^i_{(t,h)}\}_{1 \leq i \leq K+1}$ the $K+1$ (positive) eigenvalues of the matrix $(\mathbf{G}'\mathbf{G})_{(t,h)}$ and noting that the trace is also equal to the summation of the eigenvalues we get:

$$tr[\mathbf{S}_{(t,h)}] = \sigma_\epsilon^2 \, tr[(\mathbf{G}'\mathbf{G})^{-1}_{(t,h)}] = \sigma_\epsilon^2 \sum_{i=1}^{K+1} \frac{1}{\lambda^i_{(t,h)}}. \qquad [3.11]$$

To obtain [3.11], the fact that the eigenvalues of the inverse of a matrix are equal to the inverse of the eigenvalues can be used. Identity [3.9] casts light on when the worst case of large dispersion around the true value is met. It is indeed straightforward to conclude from [3.9] that $tr[\mathbf{S}_{(t,h)}]$ is lower bounded as follows:

$$tr[\mathbf{S}_{(t,h)}] > \frac{\sigma_\epsilon^2}{\lambda^{min}_{(t,h)}}. \qquad [3.12]$$

We see from [3.12] that the LSE may exhibit a large dispersion around the true parameter if the minimum eigenvalue $\lambda^{min}_{(t,h)}$ of $(\mathbf{G}'\mathbf{G})_{(t,h)}$ is small w.r.t. the observation noise variance σ_ϵ^2. This may happen when the matrix $(\mathbf{G}'\mathbf{G})_{(t,h)}$ is ill-conditioned ($\lambda^{min}_{(t,h)} \overset{>}{\to} 0$), although invertible.

3.6.2. *Geometrical interpretation of LSE*

As noted in section 3.4, the term $\mathbf{G}_{(t,h)}\hat{\boldsymbol{\theta}}_{(t,h)}$ might be viewed as the LSE $\hat{\mathbf{r}}_{(t,h)}$ of the return or measurement vector $\mathbf{r}_{(t,h)}$. In keeping with the interpretation of section 3.4, $\hat{\mathbf{r}}_{(t,h)}$

represents the linear combination of the column-regressors $\mathbf{c}^k_{(t,h)}$ (column-factors) that best fits the measurements (returns) $\mathbf{r}_{(t,h)}$ in a geometrical and Euclidian sense. Taking into account equation [3.8], we have:

$$\hat{\mathbf{r}}_{(t,h)} = \mathbf{G}_{(t,h)}\,\hat{\boldsymbol{\theta}}_{(t,h)} = \prod\nolimits_{\perp,\mathbf{G}_{(t,h)}} \mathbf{r}_{(t,h)}, \qquad [3.13]$$

where

$$\prod\nolimits_{\perp,\mathbf{G}_{(t,h)}} = (\mathbf{G}\,(\mathbf{G}'\,\mathbf{G})^{-1}\,\mathbf{G}')_{(t,h)}. \qquad [3.14]$$

The matrix $\prod_{\perp,\mathbf{G}_{(t,h)}}$ is actually the orthogonal projection matrix in the space spanned by the $K+1$ column-regressors $\mathbf{c}^k_{(t,h)}$ (column-factors). Indeed, it is easy to check that the operator $\prod_{\perp,\mathbf{G}_{(t,h)}}$ maps $\mathbf{G}_{(t,h)}$ to itself, that is $\prod_{\perp,\mathbf{G}_{(t,h)}} \mathbf{G}_{(t,h)} = \mathbf{G}_{(t,h)}$. This operator also satisfies a characteristic property of projectors, that is $\prod_{\perp,\mathbf{G}_{(t,h)}} \prod_{\perp,\mathbf{G}_{(t,h)}} = \prod_{\perp,\mathbf{G}_{(t,h)}}$. This helps us to understand that, according to [3.13], the projector nature of $\prod_{\perp,\mathbf{G}_{(t,h)}}$, $\hat{\mathbf{r}}_{(t,h)}$ is nothing but the orthogonal projection of the return-vector $\mathbf{r}_{(t,h)}$ onto the space of the column-factors $\{\mathbf{c}^k_{(t,h)}\}_{1\le k\le K+1}$. To some extent, this feature turns out to be expected: LSE $\hat{\boldsymbol{\theta}}_{(t,h)}$ builds the best geometrical and Euclidian approximation of $\mathbf{r}_{(t,h)}$ that is achieved by making its orthogonal projection in the "approximation" space, or column-factor space for the time window (t,h). This space, denoted hereafter $col\{\mathbf{G}_{(t,h)}\}$, is spanned by the columns of the matrix $\mathbf{G}_{(t,h)}$. Figure 3.1 illustrates geometrically how $\hat{\mathbf{r}}_{(t,h)}$ is obtained from $\mathbf{r}_{(t,h)}$.

Figure 3.1 also illustrates the so-called LSE orthogonality principle. Indeed, the residual $\mathbf{r}_{(t,h)} - \hat{\mathbf{r}}_{(t,h)}$ of the orthogonal projection of $\hat{\mathbf{r}}_{(t,h)}$, denoted $\tilde{\mathbf{r}}_{(t,h)}$ and also called the return-vector innovation, is geometrically orthogonal to the column-factor space. We symbolically write this orthogonality principle as follows: $\tilde{\mathbf{r}}_{(t,h)} \perp col\{\mathbf{G}_{(t,h)}\}$. This makes a lot of sense, as the innovation represents the part of the return that cannot be explained by the factors and has thus to be orthogonal to them, according to some particular "geometry". The LSE is actually based on Euclidian geometry and it is easy to prove that the Euclidian product $(\mathbf{G}'\tilde{\mathbf{r}})_{(t,h)}$ is a null vector. We first derive from [3.13] and the definition of the return innovation of the following identities:

$$\tilde{\mathbf{r}}_{(t,h)} = \mathbf{r}_{(t,h)} - \hat{\mathbf{r}}_{(t,h)} = \prod\nolimits_{\perp,\mathbf{G}_{\perp(t,h)}} \mathbf{r}_{(t,h)}, \qquad [3.15]$$

with,

$$\prod\nolimits_{\perp,\mathbf{G}_{\perp(t,h)}} = \mathbf{I} - \prod\nolimits_{\perp,\mathbf{G}_{(t,h)}}. \qquad [3.16]$$

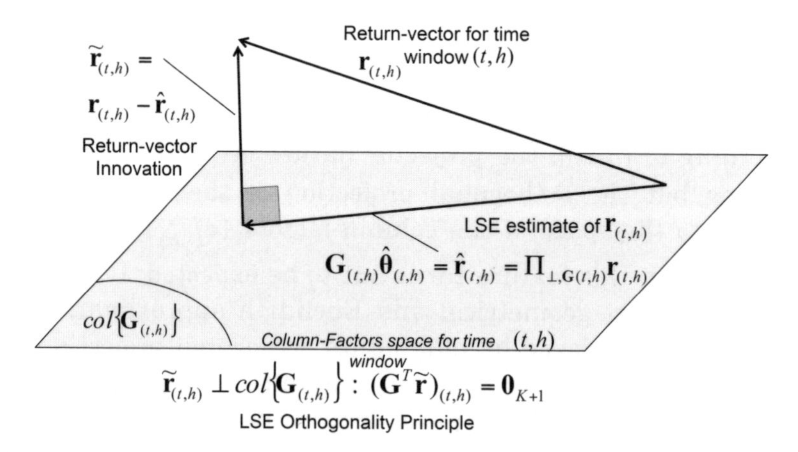

Figure 3.1. *Geometrical and Euclidian interpretation of the least square estimate (LSE) $\hat{\mathbf{r}}_{(t,h)} = \mathbf{G}_{(t,h)}\hat{\boldsymbol{\theta}}_{(t,h)}$ and the LSE orthogonality principle*

By construction, the operator $\prod_{\perp,\mathbf{G}_{\perp(t,h)}}$ involved in [3.16] happens to be the orthogonal projection matrix into the space orthogonal to the space spanned by the column-vectors of $\mathbf{G}_{(t,h)}$. Equations [3.13]–[3.16] combined with the following projection property $\mathbf{G}'_{(t,h)} \Pi_{\perp \mathbf{G}_{(t,h)}} = \mathbf{G}'_{(t,h)}$ lead to the identity chain that completes the proof:

$$(\mathbf{G}'\tilde{\mathbf{r}})_{(t,h)} = (\mathbf{G}'_{(t,h)} - \mathbf{G}'_{(t,h)} \prod_{\perp,\mathbf{G}_{(t,h)}}) \mathbf{r}_{(t,h)} = \mathbf{0}_{K+1}. \quad [3.17]$$

The LSE orthogonality principle allows us to write an orthogonal decomposition of the return-vector $\mathbf{r}_{(t,h)}$ such that, $\mathbf{r}_{(t,h)} = \hat{\mathbf{r}}_{(t,h)} + \tilde{\mathbf{r}}_{(t,h)}$ with $\hat{\mathbf{r}}_{(t,h)} = \mathbf{G}'_{(t,h)} \hat{\boldsymbol{\theta}}_{(t,h)}$ and $\tilde{\mathbf{r}}_{(t,h)} \perp \hat{\mathbf{r}}_{(t,h)}$. Pythagorean theorem subsequently yields the relationship [3.18]:

$$\|\mathbf{r}_{(t,h)}\|^2_{L^2} = \|\hat{\mathbf{r}}_{(t,h)}\|^2_{L^2} + \|\tilde{\mathbf{r}}_{(t,h)}\|^2_{L^2}. \quad [3.18]$$

It is then straightforward to evaluate the minimum value of the LSE cost function, that is $\|\hat{\mathbf{r}}_{(t,h)} - \mathbf{G}_{(t,h)} \hat{\boldsymbol{\theta}}_{(t,h)}\|^2_{L^2}$. First, because of the return vector-innovation $\tilde{\mathbf{r}}_{(t,h)}$ definition we note that the minimum value of the LSE cost function coincides with the square Euclidian norm of the return-vector innovation $\|\hat{\mathbf{r}}_{(t,h)} - \mathbf{G}_{(t,h)} \hat{\boldsymbol{\theta}}_{(t,h)}\|^2_{L^2} = \|\tilde{\mathbf{r}}_{(t,h)}\|^2_{L^2}$, see also Figure 3.1. Second, we derive from [3.18] that:

$$\|\hat{\mathbf{r}}_{(t,h)} - \mathbf{G}_{(t,h)} \hat{\boldsymbol{\theta}}_{(t,h)}\|^2_{L^2} = \|\mathbf{r}_{(t,h)}\|^2_{L^2} - \|\hat{\mathbf{r}}_{(t,h)}\|^2_{L^2}. \quad [3.19]$$

Finally, the definition of the Euclidian norm combined with equations [3.13] and [3.16] enable us to involve the projector $\Pi_{\perp,\mathbf{G}_{\perp(t,h)}}$ in the following manner:

$$\|\hat{\mathbf{r}}_{(t,h)} - \mathbf{G}_{(t,h)} \hat{\boldsymbol{\theta}}_{(t,h)}\|^2_{L^2} = \mathbf{r}'_{(t,h)} \prod_{\perp,\mathbf{G}_{\perp(t,h)}} \mathbf{r}_{(t,h)}. \quad [3.20]$$

We note from [3.20] that, for a given set of returns $\mathbf{r}_{(t,h)}$ and factor values $\mathbf{G}_{(t,h)}$ that occur across the observation window (t, h), the LSE accuracy increases when the "positive" quadratic form $\mathbf{r}'_{(t,h)} \Pi_{\perp,\mathbf{G}_{\perp(t,h)}} \mathbf{r}_{(t,h)}$ decreases. This happens when the "angle" between $\mathbf{r}_{(t,h)}$ and the approximation space $col\{\mathbf{G}_{(t,h)}\}$ is obviously small, meaning that the returns and the factors share a lot of commonalities. This conclusion was obviously an upfront common sense expectation.

3.7. Derivations of LSE solution

In keeping with the problem statement given in equation [3.6], the LSE cost function $\boldsymbol{\Omega}_{(t,h)}(\boldsymbol{\xi})$ to be minimized w.r.t. the $K+1$ dimensional real-valued vector $\boldsymbol{\xi}$ is the square Euclidian norm of the difference $\mathbf{r}_{(t,h)} - \mathbf{G}_{(t,h)}\,\boldsymbol{\xi}$. We thus have:

$$\boldsymbol{\Omega}_{(t,h)}(\boldsymbol{\xi}) = \|\mathbf{r}_{(t,h)} - \mathbf{G}_{(t,h)}\,\boldsymbol{\xi}\|_{L^2}^2 = (\mathbf{r}'\,\mathbf{r})_{(t,h)} - 2\,\boldsymbol{\xi}'\,(\mathbf{G}'\,\mathbf{r})_{(t,h)} + \boldsymbol{\xi}'\,(\mathbf{G}'\,\mathbf{G})_{(t,h)}\,\boldsymbol{\xi}. \quad [3.21]$$

$\boldsymbol{\Omega}_{(t,h)}(\boldsymbol{\xi})$ is a convex function with a unique minimum under the assumption that the positive definite matrix $\mathbf{G}'_{(t,h)}\,\mathbf{G}_{(t,h)}$ is full rank (has no vanishing eigenvalues). The vector $\boldsymbol{\xi}^{op}$ that achieves the minimum of $\boldsymbol{\Omega}_{(t,h)}(\boldsymbol{\xi})$ is thus the solution of the following equation:

$$\nabla_{\boldsymbol{\xi}}\,\boldsymbol{\Omega}_{(t,h)}(\boldsymbol{\xi})|_{\boldsymbol{\xi}=\boldsymbol{\xi}^{op}} = \mathbf{0}_{K+1}, \quad [3.22]$$

where $\nabla_{\boldsymbol{\xi}}(.)$ designates the gradient vector-operator, that is $[\nabla_{\boldsymbol{\xi}}]_k = \partial(.)/\partial\xi_k$, $1 \leq k \leq K+1$. Straightforward calculation leads to:

$$\nabla_{\boldsymbol{\xi}}\,\boldsymbol{\Omega}_{(t,h)}(\boldsymbol{\xi}) = 2\left((\mathbf{G}'\,\mathbf{G})_{(t,h)}\,\boldsymbol{\xi} - (\mathbf{G}'\,\mathbf{r})_{(t,h)}\right). \quad [3.23]$$

Equating [3.23] to zero gives finally the optimum solution [3.8]. It is interesting to grasp the behavior of the LSE cost function close to the optimum. A second-order Taylor expansion (the first-order term is zero because of the optimality condition [3.22]) gives:

$$\boldsymbol{\Omega}_{(t,h)}(\boldsymbol{\xi}) = \boldsymbol{\Omega}_{(t,h)}(\boldsymbol{\xi}^{op}) + \frac{1}{2}(\boldsymbol{\xi} - \boldsymbol{\xi}^{op})' \mathbf{H}_{\boldsymbol{\xi}^{op}_{(t,h)}} (\boldsymbol{\xi} - \boldsymbol{\xi}^{op}) + o\left(\|\boldsymbol{\xi} - \boldsymbol{\xi}^{op}\|^3\right). \quad [3.24]$$

In [3.24], $\mathbf{H}_{\boldsymbol{\xi}^{op}_{(t,h)}}$ denotes the $[K+1, K+1]$ Hessian matrix, that is $\mathbf{H}_{\boldsymbol{\xi}^{op}_{(t,h)}} = \nabla_{\boldsymbol{\xi}} \nabla'_{\boldsymbol{\xi}} \boldsymbol{\Omega}_{(t,h)}(\boldsymbol{\xi})|_{\boldsymbol{\xi}=\boldsymbol{\xi}^{op}}$. We obtain from [3.22] that $\mathbf{H}_{\boldsymbol{\xi}^{op}_{(t,h)}} = 2\,(\mathbf{G}'\,\mathbf{G})_{(t,h)}$. Substitution into [3.24] leads to:

$$\boldsymbol{\Omega}_{(t,h)}(\boldsymbol{\xi}) - \boldsymbol{\Omega}_{(t,h)}(\boldsymbol{\xi}^{op}) = (\boldsymbol{\xi} - \boldsymbol{\xi}^{op})'\,(\mathbf{G}'\,\mathbf{G})_{(t,h)}(\boldsymbol{\xi} - \boldsymbol{\xi}^{op}) + o\left(\|\boldsymbol{\xi} - \boldsymbol{\xi}^{op}\|^3\right). \quad [3.25]$$

Since $\mathbf{G}'_{(t,h)}\,\mathbf{G}_{(t,h)}$ is positive definite and full rank if $\boldsymbol{\xi} \neq \boldsymbol{\xi}^{op}$, then $\boldsymbol{\Omega}_{(t,h)}(\boldsymbol{\xi}) > \boldsymbol{\Omega}_{(t,h)}(\boldsymbol{\xi}^{op})$ close to the optimum. This confirms the convexity curvature of the cost function.

3.8. Why KF and which setup?

3.8.1. *LSE method does not provide a recursive estimate*

Coarsely (details will follow in section 3.10), both LSE and KF leverage measurements r_t that are connected to an unknown parameter $\boldsymbol{\theta}_t$, see [3.27], to find its best estimate $\hat{\boldsymbol{\theta}}_t$ based on a minimum squared distance criterion. Limited to the scope of the factor model problem, r_t is the scalar return of any asset and $\boldsymbol{\theta}_t$ is the $(K+1)$-dimensional column vector collecting the alpha and exposures (betas) of the asset (for the sake of simplicity the subscript referring to the asset has been dropped). The two other components of [3.27] have the same meaning as in the previous sections. One issue with LSE arises from the absence of recursivity of the estimate $\hat{\boldsymbol{\theta}}_t$. By recursivity, we mean the explicit updating of the new estimate $\hat{\theta}_t$ from the previous estimate $\hat{\theta}_{t-1}$. Operating the LSE across a sliding window (t, h), as described in section 3.3.2, opens a path that allows an update on the fly of the estimate but is not a recursive estimator *per se*.

Recursivity of the estimate bears virtues. The most important advantage is the tracking capability of both the estimate and its accuracy. If the main drivers of the measuring process and the evolution of the true parameter have not significantly changed from time $t-1$ to time t, then the new estimate should closely replicate the previous estimate, saving a lot of processing burden. In addition, if the new value of the true parameter and/or the new measurements depart significantly from the previous situation, then the updated estimate should reflect the change instantaneously. Real-time tracking and thus recursivity is essential in aeronautics, automatic surveillance and monitoring, car and train control, speech and image (multimedia) processing [FUK 90, BAR 93, KAY 93, DUV 94, KAY 98], intensive care medicine, econometrics and quantitative finance, etc.

The measurement (equation [3.27]), yields no recursivity of the parameter. Therefore, to build a recursive estimate, KF needs to rely on a recursive equation that drives the true parameter evolution and, moreover, that equation has to be known ahead to be included in the estimate derivation. Recursivity of the estimate will be induced by the recursivity of the true unknown parameter. This is the key difference between LSE and KF setups.

3.8.2. *The state space model and its recursive component*

This recursive equation, also called the state equation, is a discrete time version of a differential equation. It can be "natural" or "intuitive". If natural, it comes from the

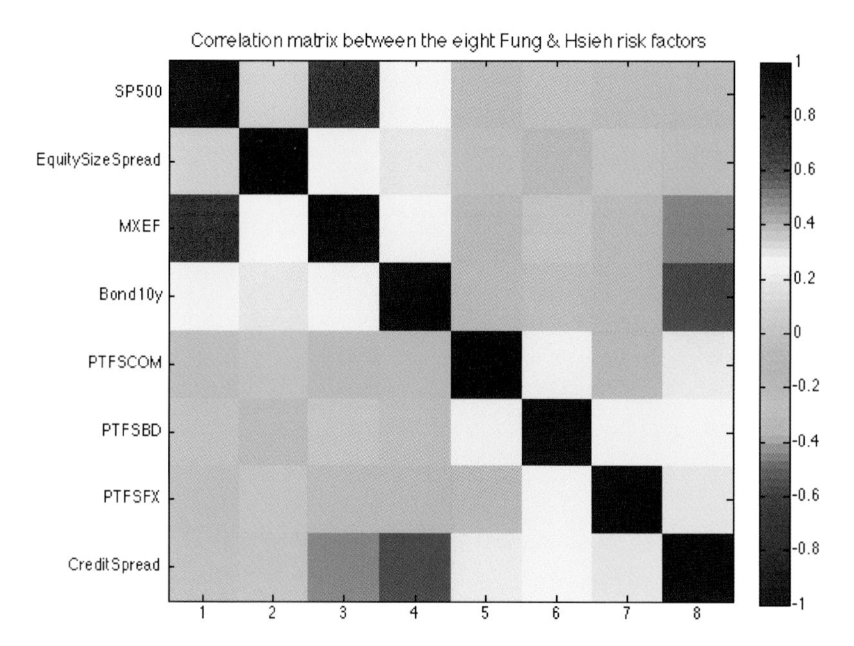

Figure 2.2.

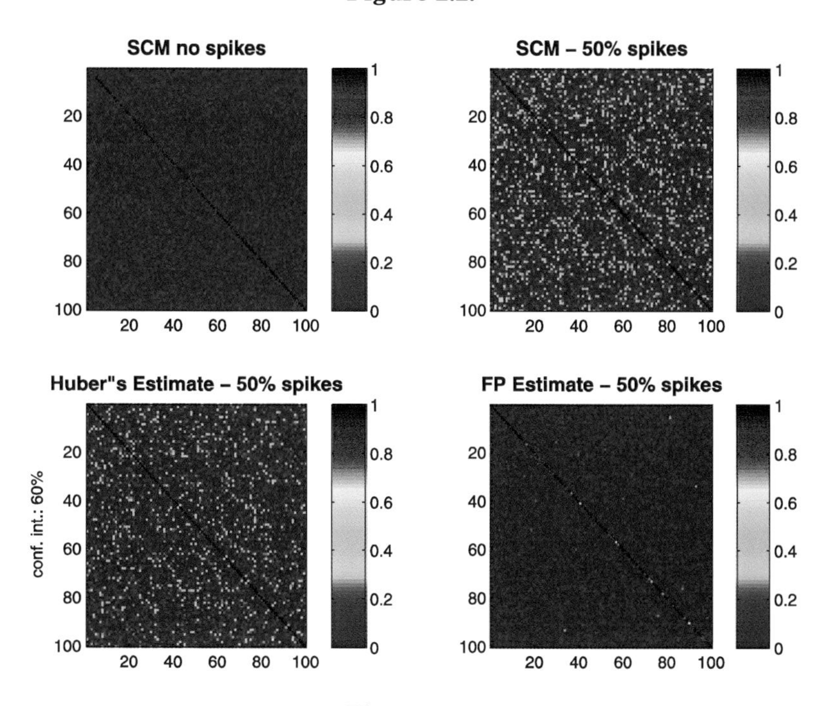

Figure 2.5.

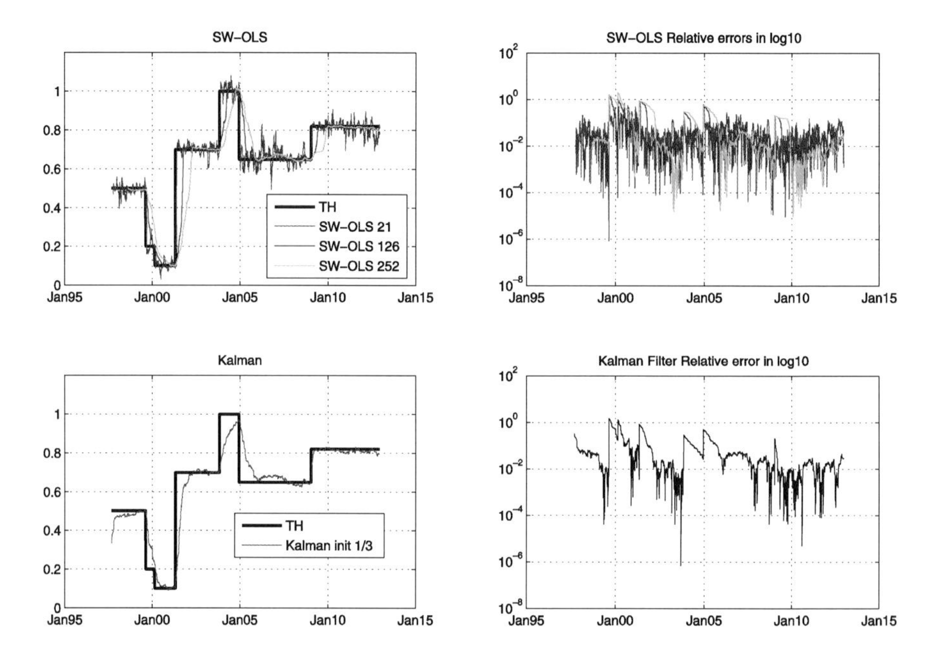

Figure 3.6.

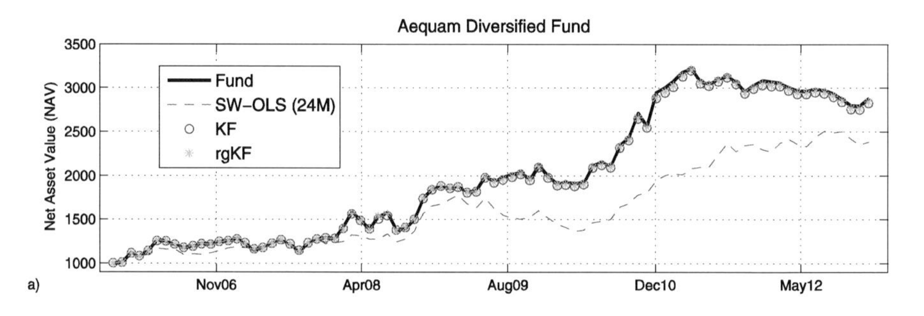

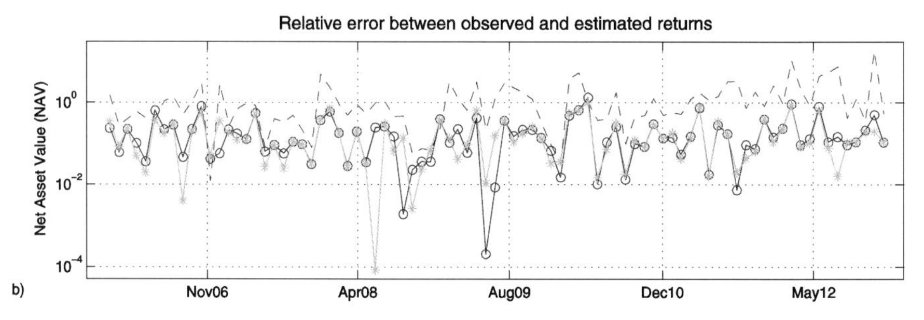

Figure 3.12.

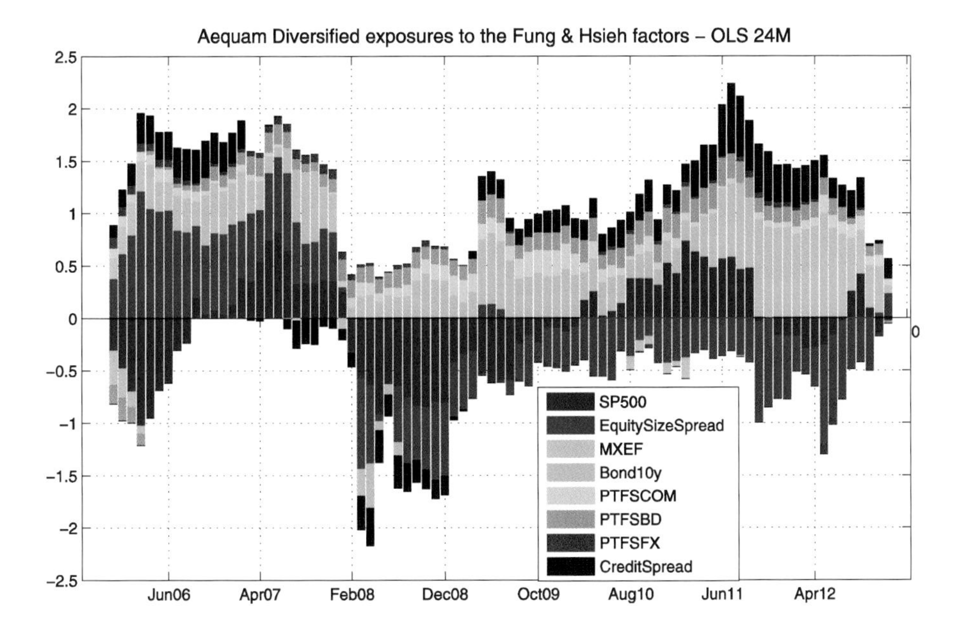

Figure 3.13.

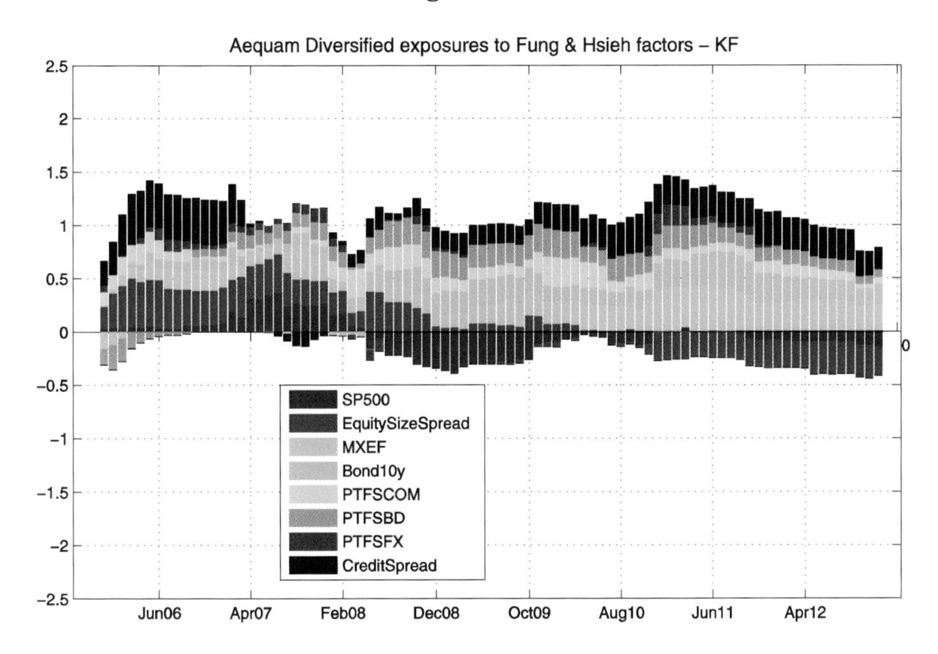

Figure 3.14.

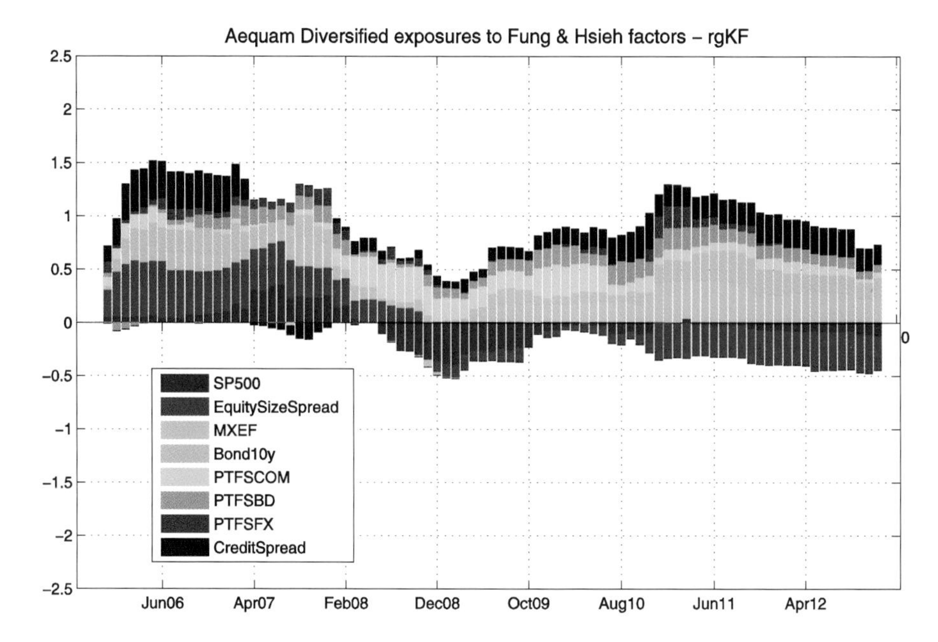

Figure 3.15.

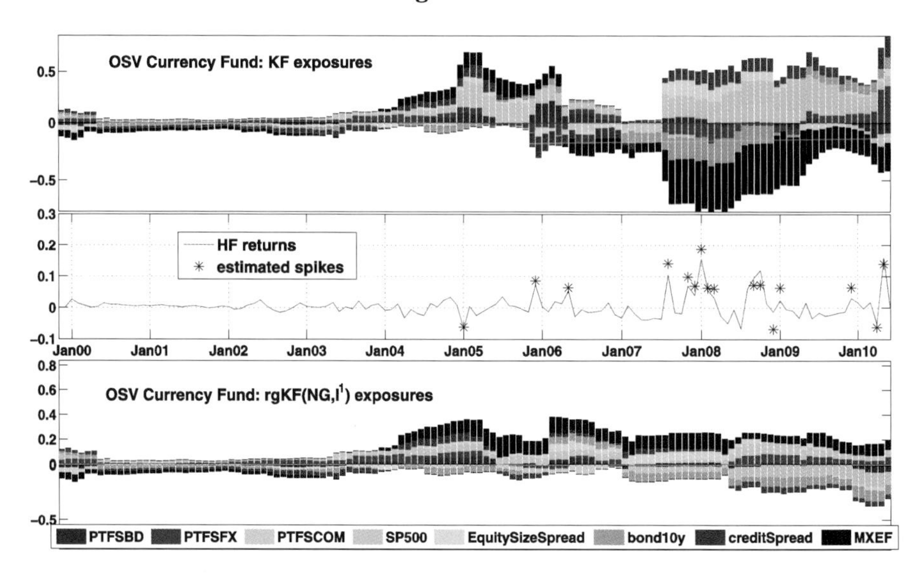

Figure 4.3.

well-known laws of physics (mechanics, thermodynamics, electromagnetism, optics, etc.) and chemistry that describe phenomena along with their key variables by differential equations. We invite the reader to check the many examples provided in [BAR 93, KAY 93, KAY 98, NOR 04]. If no particular law can be leveraged, it is still possible to postulate an "intuitive behavioral description" of the parameter evolution that is a key stone of KF.

KF state-space model / set up

Evolution of the parameters $\boldsymbol{\theta}_t$, state recursive equation:

$$\boldsymbol{\theta}_t = \boldsymbol{\theta}_{t-1} + \mathbf{w}_t, \ t \geq 1. \quad [3.26]$$

Measurement–observation equation:

$$r_t = \mathbf{g}_t' \boldsymbol{\theta}_t + \epsilon_t, \ t \geq 1. \quad [3.27]$$

No particular law of economics or finance yields a recursive equation of the alphas and exposures of any given asset. An intuitive evolution model has thus to be introduced. The simplest model leads to the fewest number of extra-parameters to be tuned and exhibits the easiest rationale to understand. The simplest and most parsimonious recursive model is given by equation [3.26]. It assumes that the new "state" of the parameter at time t is equal to the previous "state" at time $t - 1$ plus another component, the state noise $\mathbf{w}_t$, that is statistically orthogonal (uncorrelated) to the previous state. This means that, to describe the new state, we have to fully innovate w.r.t. the statistical information contained in the previous state.

3.8.3. *Parsimony and orthogonality assumptions*

Orthogonality (see also section 3.9) and parsimony are easy to achieve if we choose a zero-mean, Gaussian and white state noise with uncorrelated components, that is $\mathbb{E}[\mathbf{w}_t \mathbf{w}_t'] = \sigma_w^2 \mathbf{I}\, \delta[t_1 - t_2]$, with $\delta[0] = 1$ and $\delta[k] = 0$, $k \neq 0$, otherwise.

Consequently, the state equation depends only on one scalar parameter, that is σ_w^2, because the Gaussian probability density function depends only on the first- and second-order moments (see Appendix A1.1). The other KF model assumptions are summarized below. In keeping with the parsimony feature, both the observation noise and the initial state are also assumed Gaussian and white.

Summary of KF model assumptions

– State noise $\mathbf{w}_t \xrightarrow[\text{stationary}]{\text{white, Gaussian}} \mathcal{N}_{K+1}(\mathbf{0}, \sigma_w^2 \mathbf{I})$,

– Observation noise $\epsilon_t \xrightarrow[\text{stationary}]{\text{white, Gaussian}} \mathcal{N}(0, \sigma_\epsilon^2)$,

– Initial values $\boldsymbol{\theta}_0 \xrightarrow{\text{Gaussian}} \mathcal{N}_{K+1}(\boldsymbol{\mu}_0, \sigma_0^2 \mathbf{I})$,

– $\{\mathbf{w}_t, \epsilon_t, \boldsymbol{\theta}_0\}$ are independent and assumed known.

3.9. What are the main properties of the KF model?

3.9.1. *Self-aggregation feature*

The two key properties of the KF model formalize the main features of recursivity, parsimony and orthogonality mentioned in section 3.8. They are both based on the explicit writing of the parameter $\boldsymbol{\theta}_t$ as a function of the state noise $\mathbf{w}_t$ and the initial value $\boldsymbol{\theta}_0$. It is straightforward to derive from [3.26] the following relationship:

$$\boldsymbol{\theta}_t = \boldsymbol{\theta}_0 + \sum_{k=1}^{t} \mathbf{w}_t, t \geq 1. \qquad [3.28]$$

Equation [3.28] means, for instance, that the sole value of $\boldsymbol{\theta}_{t-1}$ self-aggregates the knowledge of the whole series $\{\boldsymbol{\theta}_0, \boldsymbol{\theta}_1, \cdots, \boldsymbol{\theta}_{t-1}\}$.

3.9.2. *Markovian property*

The self-aggregation feature is leveraged in the so-called "Markovian" property described in equation [3.29] and states that the Probability Density Function (PDF) of $\boldsymbol{\theta}_t$ conditional to its whole past is the same as its PDF conditional to its most recent value $\boldsymbol{\theta}_{t-1}$. The Markovian property expresses the recursivity feature at the probability density level of the parameter.

$$p(\boldsymbol{\theta}_t|\boldsymbol{\theta}_{t-1}, \boldsymbol{\theta}_{t-2}, \cdots, \boldsymbol{\theta}_1, \boldsymbol{\theta}_0) = p(\boldsymbol{\theta}_t|\boldsymbol{\theta}_{t-1}). \qquad [3.29]$$

Equation [3.29] can also be interpreted in terms of parsimony, because we need to propagate $\boldsymbol{\theta}_{t-1}$ only to sample the next value.

3.9.3. *Innovation property*

As w_t is white, it is statistically orthogonal to all the previous state noise samples $\{\mathrm{w}_1, \cdots, \mathrm{w}_{t-1},\}$. By "statistically orthogonal", we mean uncorrelated, that is $\mathbb{E}[\mathrm{w}_t \mathrm{w}_n'] = \mathbf{0}$, $t \neq n$, where $\mathbf{0}$ designates the zero matrix. In what follows, the symbol $\perp\!\!\!\perp$ will stand for "statistical orthogonality".

Moreover, the KF setup assumptions make w_t orthogonal to $\boldsymbol{\theta}_0$. Consequently, because of equation [3.28], w_t is orthogonal to the previous state $\boldsymbol{\theta}_{t-1}$:

$$\mathrm{w}_t \perp\!\!\!\perp \boldsymbol{\theta}_{t-1}. \qquad [3.30]$$

Equation [3.30] stands for the "innovation" (orthogonal) property: each update of the state noise brings pure innovation to the parameter evolution. As detailed in sections 3.11 and 3.12, both the Markovian and innovation properties drive the architecture of the KF.

3.10. What is the objective of KF?

Given the measurements up to time t, $\{r_k, 1 \leq k \leq t\}$ AND the state-space setup (model) described by equations [3.26] and [3.27], KF aims at deriving a *recursive and linear* (regarding linearity, see the second and third notes hereafter) estimate $\hat{\boldsymbol{\theta}}_{t|t}$ of the unobservable state or parameter $\boldsymbol{\theta}_t$, $t \geq 1$ based on the minimization of the mean square error (MSE). The second subscript q of $\hat{\boldsymbol{\theta}}_{t|q}$ stands for the end of the time window of the observation space $R_1^q = \{r_k, 1 \leq k \leq q\}$ used to estimate $\boldsymbol{\theta}_t$, $t \geq 1$. If the problem is limited to KF, then $q = t$. For Kalman prediction and smoothing, not addressed in this book, we would have $q < t$ and $q > t$ respectively.

KF objectives:

$$\overbrace{R_1^t = \{r_k, 1 \leq k \leq t\} \text{ AND State-Space Model}}^{\text{given}} \xrightarrow{\text{KalmanFiltering}} \text{Recursive}(\hat{\boldsymbol{\theta}}_{t|t}), t \geq 1, \quad [3.31]$$

Based on minimum mean square error (MMSE) optimality criterion:

$$\hat{\boldsymbol{\theta}}_{t|t} = Arg\left\{min_{\boldsymbol{\xi}_{|t} \in R_1^t}\left\{\mathbb{E}\left[\|\boldsymbol{\theta}_t - \boldsymbol{\xi}_{|t}\|^2\right]\right\}\right\}. \quad [3.32]$$

In equation [3.32], $\boldsymbol{\xi}_{|t}$ designates any element of the observation space R_1^t spanned by the random variables $\{r_1, \cdots, r_t\}$, see section 3.14 for the geometrical and Euclidian interpretation of MMSE criterion.

NOTE 3.2.– The recursivity of the estimate is naturally induced by the recursivity of the state equation [3.26] and the consecutive Markovian property [3.29].

NOTE 3.3.– For the sake of simplicity and for the benefit of newcomers, we have positioned the objectives of KF as a recursive MMSE estimate, limiting thus the optimality search to an intuitive "distance-like" or geometrical-like second-order criterion (see section 3.14). The geometrical path skips the probabilistic or Bayesian presentation, based on conditional probability densities. Probabilistic derivation requires a lengthy introduction of maximum *a posteriori* methods and detailed properties of normal conditional PDFs, at the expense of the intuitive and quick understanding of KF structure, behavior and performance.

NOTE 3.4.– In the probabilistic (Bayesian) approach and under Gaussian assumptions, linearity of the estimate w.r.t. both the measurements and the previous estimate is a consequence of the optimization process. The MMSE-based presentation undertaken in this chapter needs to specify the structure of the estimate. It is limited hereafter to linear operators in order to lead exactly to the same solution as the probabilistic derivation.

3.11. How does the KF work (for users and programmers)?

3.11.1. *Algorithm summary*

Resolution of the problem stated in section 3.10 is detailed in section 3.14. Hereafter, we describe and interpret the recursive structure of the KF solution that should be well understood by both KF users and programmers. One *full recursive KF step* from time $t-1$ to time t, with $t \leq 1$, is twofold. Indeed, as displayed in Figure 3.2, the KF full step updates, on the one hand, the (alphas and betas), from $\hat{\theta}_{t-1|t-1}$

to $\hat{\boldsymbol{\theta}}_{t|t}$, and, on the other hand, the associated KF "accuracy" matrix, from $\mathbf{S}_{t-1|t-1}$ to $\mathbf{S}_{t|t}$, with $\mathbf{S}_{n|q} = \mathbb{E}[\tilde{\boldsymbol{\theta}}_{n|q}\tilde{\boldsymbol{\theta}}'_{n|q}]$. $\tilde{\boldsymbol{\theta}}_{n|q}$ is called the $KF(n/q)$ innovation, $\tilde{\boldsymbol{\theta}}_{n|q} = \boldsymbol{\theta}_n - \hat{\boldsymbol{\theta}}_{n|q}$. The accuracy matrix is nothing but the covariance matrix of the current KF innovation. The decrease of its dynamic range tracks the accuracy improvement and thus the convergence of the KF estimate closer to the targeted parameter $\boldsymbol{\theta}_t$.

1) Initialization $t = 0$

KF is initialized in the following manner:

$$\hat{\boldsymbol{\theta}}_{0|0} = \boldsymbol{\mu}_0;\ \mathbf{S}_{0|0} = \sigma_0^2\,\mathbf{I};\ (\sigma_0^2, \boldsymbol{\mu}_0) \text{ known} \qquad [3.33]$$

From time $t-1$ to time t, with $t \leq 1$: full KF recursive processing breaks into (1) prediction (also called *a priori* update) and (2) correction (also called *a posteriori* update), see Figure 3.2.

2) Prediction, from $(t-1|t-1)$ to $(t|t-1)$.

Prediction is the sole genuine recursive step of KF induced by the recursive (Markovian) state equation of the model (see Figure 3.2). The prediction step relies only on the confidence in the recursive state-space KF model given by equation [3.26]. Prediction updates are split into prediction filtering that computes the new state estimate $\hat{\boldsymbol{\theta}}_{t|t-1}$ from the previous state estimate $\hat{\boldsymbol{\theta}}_{t-1|t-1}$, see [3.34], and prediction accuracy processing, see [3.35], that refreshes of the accuracy matrix.

Prediction filtering

$$t \geq 1 \qquad \hat{\boldsymbol{\theta}}_{t|t-1} = \hat{\boldsymbol{\theta}}_{t-1|t-1} \qquad [3.34]$$

Prediction accuracy processing

$$\mathbf{S}_{t|t-1} = \mathbf{S}_{t-1|t-1} + \sigma_w^2\,\mathbf{I} \qquad [3.35]$$

3) Correction, from $(t|t-1)$ to $(t|t)$.

Correction is triggered by the capture of the updated measurement r_t (see Figure 3.2). After the calculation of the so-called Kalman gain, see [3.36], the correction filtering sub-step computes $\hat{\theta}_{t|t}$, see [3.37]. The correction accuracy processing evaluates $\mathbf{S}_{t|t}$, see [3.38].

Correction filtering $t \geq 1$

Kalman gain

$$\mathbf{z}_t = \mathbf{S}_{t|t-1}\,\mathbf{g}_t\left(\mathbf{g}_t'\,\mathbf{S}_{t|t-1}\,\mathbf{g}_t + \sigma_\epsilon^2\right)^{-1} \quad [3.36]$$

Correction update

$$\hat{\theta}_{t|t} = \hat{\theta}_{t|t-1} + \mathbf{z}_t\left(r_t - \mathbf{g}_t'\,\hat{\theta}_{t|t-1}\right) \quad [3.37]$$

Correction accuracy processing

$$\mathbf{S}_{t|t} = \left(\mathbf{I} - \mathbf{z}_t\,\mathbf{g}_t'\right)\mathbf{S}_{t|t-1} \quad [3.38]$$

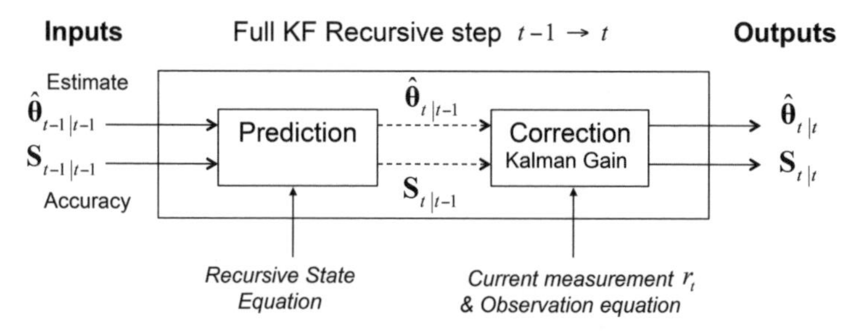

Figure 3.2. *Inputs / outputs schematic of Kalman filtering prediction and correction steps*

As displayed in equation [3.37], the "Kalman gain" $\mathbf{z}_t$ defined by [3.36] helps us to balance, on the one hand, the weight of the recursive model reflected in $\hat{\theta}_{t|t-1}$ and, on the other hand, the weight of the measurement current value and the observation equation [3.27].

Figure 3.2 displays the schematic of the full KF recursive step that is split into prediction and correction processing, as per equations [3.34]–[3.38]. The KF schematic emphasizes the key roles of the state recursive equation [3.26] and the observation [3.27] model in each of the two steps.

3.11.2. *Initialization of the KF recursive equations*

Similar to several algorithms where initial values have to be chosen, there is a trade-off to make between the estimation accuracy and the convergence speed of the algorithm. Depending of the initial values, recursive algorithms will converge more or less quickly toward the true values. There is no universal rule, only some "good sense rules".

As stated in equation [3.33], the parameters of the space-state model have to be initialized. The initial values of the vector $\hat{\boldsymbol{\theta}}_{0|0}$ have to be set regarding the knowledge of the problem and what represents this vector. In the case of factor model, $\boldsymbol{\theta}$ accounts for the exposures of the portfolio to some market factors, so that their values are mostly comprised between -1 and $+1$. Initializing $\hat{\boldsymbol{\theta}}_{0|0}$ with zeros is therefore very plausible.

The initial value of the variance of the estimation error is also linked to $\boldsymbol{\theta}$ and should not be too small: if so, then the Kalman gain will also be quite small, and regarding equation [3.37], the correction value of $\hat{\boldsymbol{\theta}}_{t|t}$ will remain close to the prediction value $\hat{\boldsymbol{\theta}}_{t|t-1}$. Then, if the true values of $\boldsymbol{\theta}$ are not close to the initial value, the algorithm will never reach these true values as it will always propose some values close to the initial values.

In addition, if σ_0^2 is initialized to a value which is too large, then the algorithm will converge slower toward the true value than if σ_0^2 were set to a more convenient value.

In our case, we may set σ_0^2 to 1 or 1.5 as from 0 we wish to reach the lowest and the highest bound (–1 and +1) of the range of potential values.

The two other parameters σ_w^2 and σ_ϵ^2 of the space-state model are supposed to be known. The variance σ_ϵ^2 of the observation equation represents the magnitude of the error made on the model so that it may be set as a fraction of the variance of the observed returns. The variance σ_w^2 impacts the prediction accuracy processing [3.35] and has also to be set in concordance with the amplitude and assumed variability of the state variable.

3.12. Interpretation of the KF updates

All the prediction and correction updating equations will be derived in section 3.14 based on a simple "geometric and Euclidian approach". Hereafter, we give intuitive interpretations of each of them. These qualitative rationales turn out to be very useful when programming and/or using KF, especially to monitor the performance and tune the KF inputs.

3.12.1. *Prediction filtering, equation [3.34]*

Equation [3.34] predicts the estimate $\hat{\boldsymbol{\theta}}_{t|t-1}$ based on $\hat{\boldsymbol{\theta}}_{t-1|t-1}$. According to the notations introduced in section 3.10, both estimates $\hat{\boldsymbol{\theta}}_{t|t-1}$ and $\hat{\boldsymbol{\theta}}_{t-1|t-1}$ are based on the same measurement (returns) samples contained in the observation set, up to time $t-1$, that is $R_1^{t-1} = \{r_k, 1 \leq k \leq t-1\}$. According to the state recursive equation [3.26], the only difference between $\boldsymbol{\theta}_t$ and $\boldsymbol{\theta}_{t-1}$ is the state noise w_t that is statistically orthogonal (uncorrelated with) to $\boldsymbol{\theta}_{t-1}$, because of the "innovation property" (see section 3.9). Moreover, w_t is not correlated to the observation noise ϵ_t, by assumption. This means that $\mathbf{w}_t$ is statistically orthogonal to the measurement set R_1^{t-1}. Therefore, any of the samples contained in R_1^{t-1} cannot contribute to an estimation of $\mathbf{w}_t$. There is thus no rationale to change the estimate $\hat{\boldsymbol{\theta}}_{t-1|t-1}$ when switching to $\hat{\boldsymbol{\theta}}_{t|t-1}$, as expressed in equation [3.34].

3.12.2. *Prediction accuracy processing, equation [3.35]*

As defined at the beginning of this section, the matrix $\mathbf{S}_{n|q}$ is a covariance matrix and, as such, is a non-negative definite matrix (all its eigenvalues are greater or equal to zero). For the sake of fluidity of the present interpretation, we assume that $\mathbf{S}_{n|q}$ is full rank (none of its eigenvalues is vanishing). Consequently, its inverse $\mathbf{S}_{n|q}^{-1}$ exists and is positive definite. Both their traces are positive, and thus the trace of $\mathbf{S}_{n|q}^{-1}$ denoted $tr[\mathbf{S}_{n|q}^{-1}]$ can be used, in the context of KF, as a positive measure or magnitude of the updates accuracy. Indeed, the larger the trace $tr[\mathbf{S}_{n|q}^{-1}]$, the "smaller" the error, and the larger the accuracy. If the matrix $\mathbf{S}_{n|q}$ is zero (the error is zero), the accuracy degree becomes "infinite" which is consistent with common sense. We denote $\lambda_{n|q}^k$, $1 \leq k \leq K+1$ the eigenvalues of $\mathbf{S}_{n|q}$ that are all assumed to be strictly positive for the sake of simplicity of the current interpretation, that is $\lambda_{n|q}^k > 0$. Taking into account that the eigenvalues of the inverse of a matrix are equal to the inverse of the eigenvalues of the matrix and noting that the trace is also the summation of the eigenvalues, it is then straightforward to derive from [3.35] and the definition of the accuracy degree the following relationship:

$$tr[\mathbf{S}_{t|t-1}^{-1}] - tr[\mathbf{S}_{t-1|t-1}^{-1}] = -\sum_{k=1}^{K+1} \frac{1}{\lambda_{t-1|t-1}^k} \frac{\sigma_w^2}{(\lambda_{t-1|t-1}^k + \sigma_w^2)}. \quad [3.39]$$

Equations [3.34] and [3.39] mean that the prediction step decreases the "accuracy", because (obviously)

$tr[\mathbf{S}_{t|t-1}^{-1}] - tr[\mathbf{S}_{t-1|t-1}^{-1}] \leq 0$. Inherently, this is not surprising: the state noise $\mathbf{w}_t$ that is part of the recursive model [3.26] is statistically orthogonal to R_1^{t-1}. Consequently, no change of the estimate takes place (equation [3.34]) in the prediction step, while the previous state $\boldsymbol{\theta}_{t-1}$ undergoes the additive state noise (equation [3.26]). Accuracy will improve only across the correction step that fully benefits from the new return measurement r_t.

3.12.3. *Correction filtering equations [3.36]–[3.37]*

We start the correction updates interpretation by noting that, in equation [3.37], the term $\mathbf{g}_t' \hat{\boldsymbol{\theta}}_{t-1}$ is a genuine estimate $\hat{r}_{t|t-1}$ of the current measurement r_t based on its past samples contained in R_1^{t-1}. Therefore, the difference $r_t - \mathbf{g}_t' \hat{\boldsymbol{\theta}}_{t-1}$ equal to $r_t - \hat{r}_{t|t-1}$ stands for an error signal, denoted $\tilde{r}_{t|t-1}$, and called, in what follows, the current measurement (return) innovation. Moreover, the difference $r_t - \mathbf{g}_t' \hat{\boldsymbol{\theta}}_{t-1}$ might also be viewed as an estimate $\hat{\epsilon}_{t|t-1}$ of the observation noise ϵ_t. Finally, we have the following identities chain that turns out to be truly meaningful to understand the different aspects of equation [3.37]:

$$r_t - \mathbf{g}_t' \hat{\boldsymbol{\theta}}_{t-1} = \tilde{r}_{t|t-1} = r_t - \hat{r}_{t|t-1} = \hat{\epsilon}_{t|t-1}. \qquad [3.40]$$

Substitution of the first of the above identities in [3.37] yields the following simple form of the correction KF updates:

$$\hat{\boldsymbol{\theta}}_{t|t} = \hat{\boldsymbol{\theta}}_{t|t-1} + \mathbf{z}_t \, \tilde{r}_{t|t-1}. \qquad [3.41]$$

Equation [3.37] and its more concise version [3.41] clearly express that the correction update, that is $t|t-1 \rightarrow t|t$, balances the weight of predicted estimate $\hat{\boldsymbol{\theta}}_{t|t-1}$ (based on the prediction filtering) and the weight of the measurement (return) estimation error (innovation) $\tilde{r}_{t|t-1}$ via the Kalman gain $\mathbf{z}_t$. According to equation [3.36], for a given covariance matrix $\mathbf{S}_{t|t-1}$, when the observation noise variance σ_ϵ^2 increases, the Kalman gain norm:

$$\|\mathbf{z}_t\| = \frac{\|\mathbf{S}_{t|t-1}\,\mathbf{g}_t\|}{\mathbf{g}_t'\,\mathbf{S}_{t|t-1}\,\mathbf{g}_t + \sigma_\epsilon^2}, \qquad [3.42]$$

decreases, subsequently reducing the weight of the measurement (return) innovation $\tilde{r}_{t|t-1}$ in the final update. This self-regulation of the correction is actually expected. When the observation noise is large, the measurement estimate is not reliable and the associated error should be marginalized, in the overall updating process to prevent further contamination of the current estimate. The Kalman gain norm is also drastically reduced when, for a given σ_ϵ^2, the covariance matrix $\mathbf{S}_{t|t-1}$ becomes "small", like close to the zero matrix. In this case, the Kalman gain vanishes since the filter has converged close to the solution and any new measurement will not significantly improve the estimate. In this latter situation, the KF reliability is mostly put on the predicted component $\hat{\boldsymbol{\theta}}_{t|t-1}$.

Conversely, when the variance σ_ϵ^2 of the observation noise decreases, the measurement component of the update [3.37] becomes more reliable and should thus be more emphasized which is the case since the Kaman gain norm [3.42] subsequently increases.

3.12.4. *Correction accuracy processing, equation [3.38]*

To interpret the accuracy update contained in equation [3.38], it is appropriate to replace the Kalman gain by its explicit form given by equation [3.36]. In doing so, equation [3.38] becomes:

$$\mathbf{S}_{t|t} = \mathbf{S}_{t|t-1} - \frac{\mathbf{S}_{t|t-1}\,\tilde{\mathbf{g}}_t\,\tilde{\mathbf{g}}_t'\,\mathbf{S}_{t|t-1}}{1 + \tilde{\mathbf{g}}_t'\,\mathbf{S}_{t|t-1}\,\tilde{\mathbf{g}}_t}, \qquad [3.43]$$

with $\tilde{\mathbf{g}}_t = \mathbf{g}_t/\sigma_\epsilon$. Moreover (still assuming that the matrix $\mathbf{S}_{t|t-1}$ is full rank for the sake of simplicity), the matrix inversion lemma (see Appendix 3.16) yields:

$$(\mathbf{S}_{t|t-1}^{-1} + \tilde{\mathbf{g}}_t\,\tilde{\mathbf{g}}_t')^{-1} = \mathbf{S}_{t|t-1} - \frac{\mathbf{S}_{t|t-1}\,\tilde{\mathbf{g}}_t\,\tilde{\mathbf{g}}_t'\,\mathbf{S}_{t|t-1}}{1 + \tilde{\mathbf{g}}_t'\,\mathbf{S}_{t|t-1}\,\tilde{\mathbf{g}}_t}. \qquad [3.44]$$

Combining [3.43], [3.44] and [3.45] leads to the following chain of updates:

$$\mathbf{S}_{t|t}^{-1} = \mathbf{S}_{t|t-1}^{-1} + \frac{\mathbf{g}_t\,\mathbf{g}_t'}{\sigma_\epsilon^2} = (\mathbf{S}_{t-1|t-1} + \sigma_w^2\,\mathbf{I})^{-1} + \frac{\mathbf{g}_t\,\mathbf{g}_t'}{\sigma_\epsilon^2}. \qquad [3.45]$$

Equation [3.45] is easy to interpret in terms of degree of accuracy. As a matter of fact, equation [3.38] and consequently [3.45] mean that the correction step always improves the accuracy w.r.t. of the prediction step, because [3.45] implies that:

$$tr[\mathbf{S}_{t|t}^{-1}] = tr[\mathbf{S}_{t|t-1}^{-1}] + \frac{\mathbf{g}_t'\,\mathbf{g}_t}{\sigma_\epsilon^2}. \qquad [3.46]$$

Recursion [3.46] clearly means that $tr[\mathbf{S}_{t|t}^{-1}] > tr[\mathbf{S}_{t|t-1}^{-1}]$: the correction step always increases degree of accuracy, as much as the quantity $\mathbf{g}_t'\,\mathbf{g}_t/\sigma_\epsilon^2$ is large. This latter entity that may be viewed as "factor-to-observation noise ratio" of the measurement becomes marginal if the variance σ_ϵ^2 of the observation noise drastically increases, thus limiting significantly the accuracy gain. Another virtue of the update [3.45] arises from the "numerical" propagation of the positive-definiteness of the inverse of the covariance matrix $\mathbf{S}_{t|t-1}^{-1}$. Indeed, we derive from [3.45] that, whatever is the non-zero vector $\mathbf{u}$:

$$\mathbf{u}'\,\mathbf{S}_{t|t}^{-1}\,\mathbf{u} = \mathbf{u}'\,\mathbf{S}_{t|t-1}^{-1}\,\mathbf{u} + \frac{(\mathbf{u}'\,\mathbf{g}_t)^2}{\sigma_\epsilon^2}. \qquad [3.47]$$

Identity [3.47] leads to:

$$\forall \mathbf{u} \neq 0, \qquad \mathbf{u}' \, \mathbf{S}_{t|t-1}^{-1} \, \mathbf{u} > 0 \Rightarrow \mathbf{u}' \, \mathbf{S}_{t|t}^{-1} \, \mathbf{u} > 0. \qquad [3.48]$$

3.13. Practice

This section is the application part of this chapter. We will study three different cases:

(1) a toy example, where synthetic observed returns are generated like a linear combination of two observed and known factors, the S&P500 Index and the US 10Y-Bond,

(2) a one-factor model used to hedge accurately the systematic risk of a given portfolio (section 3.13.2) and then

(3) a multi-factor model used in order to analyze and decompose the exposition of trend-follower hedge funds to several market risks as defined by Fung and Hsieh [FUN 01] (section 3.13.3).

3.13.1. *Comparison of the estimation methods on synthetic data*

3.13.1.1. *Data description*

In this example, we generate synthetic observed returns r_t on the period September 10 1997/December 19 2012 (around 15 years long with $T = 3,986$ daily numbers) using two market factors and given theoretical parameters $\left[w_t \; (1 - w_t)\right]'$ accounting for the betas. The two market factors are:

– the daily returns of the first Generic Future on the mini-S&P500 Index, denoted as r^{sp500},

– the daily returns of the first Generic Future on the US 10Y Treasury note, denoted as $r^{tnote10y}$, and the resulting synthetic returns are written for $t = 1, \cdots, T$:

$$r_t = w_t \, r_t^{sp500} + (1 - w_t) \, r_t^{tnote10y}. \qquad [3.49]$$

Over the whole period, the means and standard deviations of the two market factors are reported in Table 3.1.

	Mean	Standard deviation
Future on S&P500 Index	$0.135.10^{-3}$	0.0133
Future on US 10Y T. Note	$0.269.10^{-3}$	0.0058

Table 3.1. *Mean and standard deviation of the time series of the two market factor used in the toy example*

Two different cases are considered: (1) w_t constant over time and (2) w_t piecewise constant over time. For each case, we will compare the estimation results of sliding window ordinary least squares (SW-OLS) and KF.

– First case: w_t is constant over time; no observation error

Figure 3.3 shows the evolution (base 1,000 at the beginning of the period) of the time series for the two selected market factors and the resulting synthetic data for $w_t = 0.3$, $\forall t$, that is when w_t is constant over time.

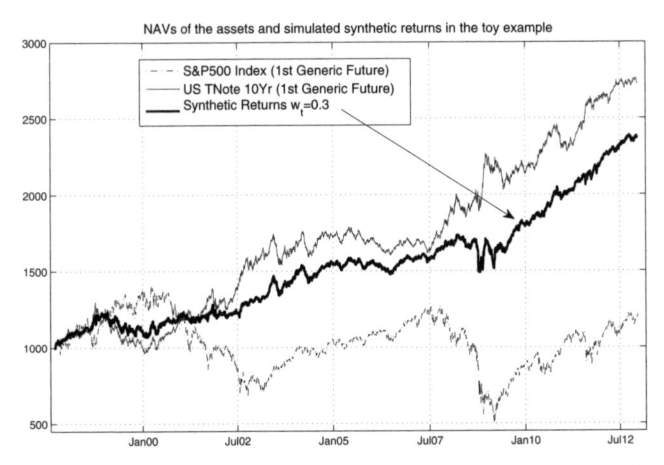

Figure 3.3. *Time series of the two selected assets in the toy example and the resulting synthetic data obtained with a constant parameter* $w_t = 0.3$

A state equation is added to [3.49] to apply a KF. The model is therefore written as:

$$r_t = a_t + w_t \, r_t^{sp500} + (1 - w_t) \, r_t^{tnote10y},$$
$$w_t = w_{t-1} + \epsilon_t, \qquad [3.50]$$

where $\epsilon \sim \mathcal{N}(0, \sigma_\epsilon^2 \mathbf{I})$ with $\sigma_\epsilon = 10^{-1}$.

In that (unrealistic) particular case, the ordinary least squares (OLS)-family methods are the optimal methods. In Figure 3.4, we have reported the estimation results for a SW-OLS (using a 252-long window) and a KF, starting with two different initial values. As expected, SW-OLS returns exactly the theoretical value, whereas the KF estimate converges to the theoretical value after a period of almost 3 years of data, depending also on the initial value.

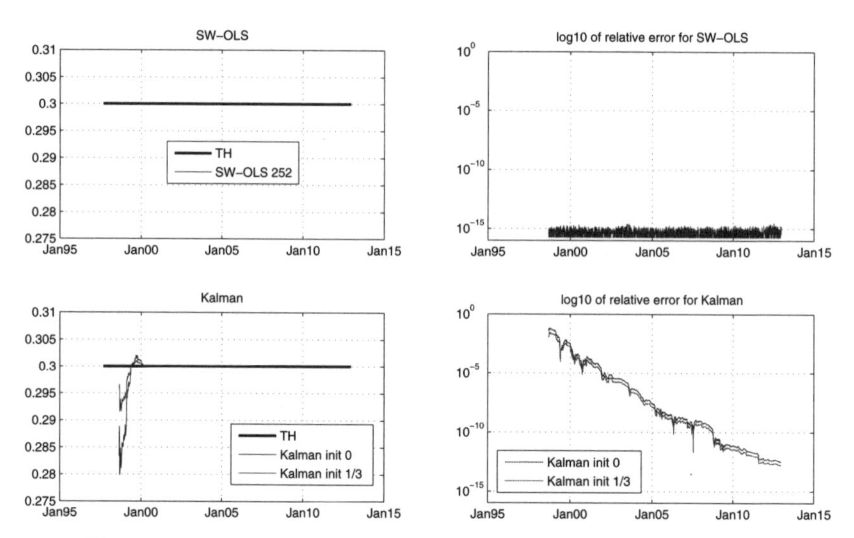

Figure 3.4. *SW-OLS and Kalman filter estimation results of the constant parameter (w_t = 0.3) in the toy example*

If we ever add a Gaussian random noise even with a small variance when generating the returns in [3.49], then the results differ and the KF estimates yield to lower MMSE than SW-OLS for small window lengths. To illustrate this result, we have run $N_{simu} = 1,000$ simulations:

For each case $i = 1, \cdots, N_{simu}$:

a) Generate synthetic noisy returns as:

$$r_t = w_t \, r_t^{sp500} + (1 - w_t) \, r_t^{tnote10y} + \sigma^2 \, \xi_t, \qquad [3.51]$$

with $\sigma = 10^{-3}$ and $\xi \sim \mathcal{N}(0, 1)$.

b) SW-OLS:

- for each time t = h + 1 to T, estimate $\hat{w}_t$ with SW-OLS given two different window lengths, h = 21 and h = 252;

- compute the root mean square error (RMSE): $\text{rmse}(i) = \sqrt{\frac{1}{T-h} \sum_{k=h+1}^{T} (\hat{w}_k - w_k)^2}$.

c) Kalman filter:

- estimate $\hat{w}_t$ using the KF recursive equations;

- compute the RMSE: $\text{rmse}(i) = \sqrt{\frac{1}{T} \sum_{k=1}^{T} (\hat{w}_k - w_k)^2}$

Note that we could remove the h-first estimated values.

At the end, we compare in Figure 3.5 the distributions of the RMSEs through a boxplot of the N_{simu} values of the RMSEs.

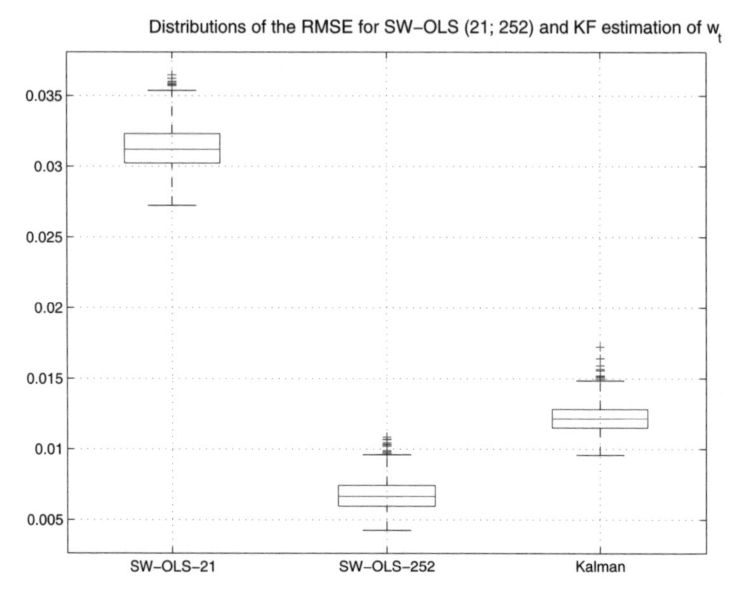

Figure 3.5. *Comparison of the RMSEs for SW-OLS ($h = 21, 252$) and Kalman filter for the synthetic noisy returns of equation [3.51] having a constant parameter $w_t = 0.3$*

– Second case: w_t is piecewise constant

Let us now consider that w_t is piecewise constant:

$$w_t = \begin{cases} 0.5 = & t \in [1, 504[, \text{or from 10-Sep-1997 to 16-Aug-1999}, \\ 0.2 = & t \in [505, 640[, \text{or from 17-Aug-1999 to 22-Feb-2000}, \\ 0.1 = & t \in [641, 950[, \text{or from 23-Feb-2000 to 1-May-2001}, \\ 0.7 = & t \in [951, 1600[, \text{or from 2-May-2001 to 28-Oct-2003}, \\ 1 = & t \in [1601, 1891[, \text{or from 29-Oct-2003 to 8-Dec-2004}, \\ 0.65 = & t \in [1892, 2957[, \text{or from 9-Dec-2004 to 8-Jan-2009}, \\ 0.82 = & t \in [2958, T], \text{or from 9-Jan-2009 to 19-Dec-2012} \end{cases}$$

[3.52]

The resulting estimations of the piecewise constant parameter are given in Figure 3.6. Overall, SW-OLS and KF react quite well to the changes in parameter w_t. It should be noted that the variance of the estimation is lower for the KF than for SW-OLS, especially when SW-OLS is used with small window lengths. Moreover and as expected, the SW-OLS estimates are more lagged than the KF estimates.

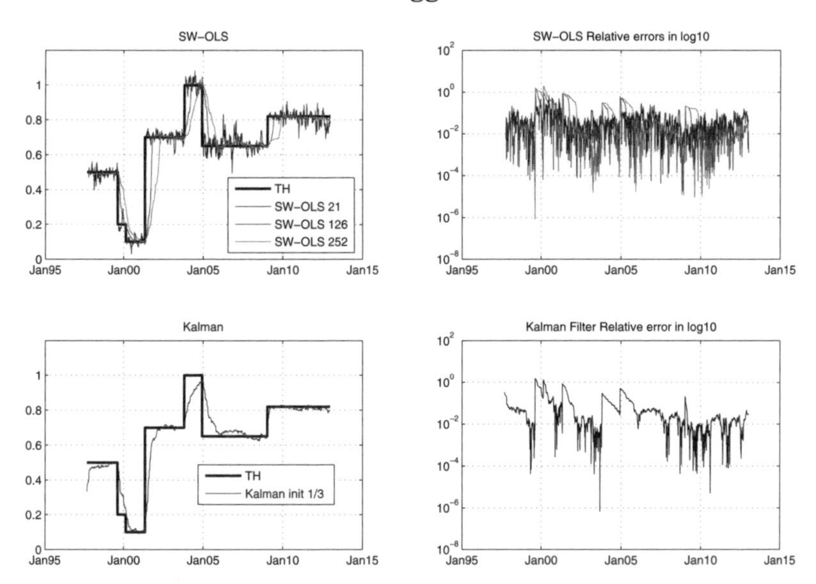

Figure 3.6. *Comparison of the SW-OLS (h = 21, 126, 252) and KF for the synthetic and noisy returns of equation* [3.51] *having a piecewise constant parameter w_t as defined in [3.52] (see color plate section)*

In practice, portfolio managers do not turn their portfolio so sharply. This case is extreme but highlights the tracking capabilities of the KF in a dynamic environment. It confirms the growing interest in such filtering techniques to dynamically hedge a portfolio or to estimate the nonlinear and dynamic exposures in hedge returns.

The next example is focused on a single-factor model applied to hedge dynamically a truly managed long-only Indian equity portfolio. The objective is to accurately estimate the exposure of the managed portfolio to its reference index, the Nifty Index.

3.13.2. *Market risk hedging given a single-factor model*

Let us assume in this section that the observed returns r_t are the returns of a long-only equity portfolio. The portfolio manager selects a subset of stocks in which to invest and define an allocation either based on his/her own market feeling or resulting from an investment committee.

A long-only equity portfolio is usually constructed and managed in order to over-perform a given stock index that often results in a strong historical correlation between the returns of the portfolio and the stock index itself. What is expected from such an investment is to deliver a performance which is larger than the performance given by the stock index, and this may be due to the manager's skills to select stocks which will deliver some α over the performance of the benchmark.

Given this setup, we can observe that the market risk is not hedged at all. The long-only strategy is not expected to be hedged against any risk arising in the market, but only to deliver a better performance than a given benchmark.

In the case where the market risk has to be hedged, we have to answer to "how much the portfolio is market-dependent?"

As explained by the factor model, if there is a single market risk (the benchmark stock index), then a single-factor model may be used to link the portfolio returns to the market returns:

$$r_t = \alpha_t + \beta_t f_t + \epsilon_t,$$

where α_t is the time-varying "alpha", β_t is the time-varying exposure of the portfolio to the single factor f_t, and ϵ_t has the properties described in the previous section.

In this application, the returns of the long-only equity portfolio are the returns of an Indian equity portfolio managed by an asset manager in India. The manager selects a subset of equity stock according to a discretionary approach based on an in-depth analysis of the fundamental numbers related to the balance sheet of the companies and on a set of specific financial ratios. The selected names are mainly growth companies for which the manager has a strong belief that these companies are under-valued and have a high level of future growth. Each name is kept in the portfolio for a given period, from 1 month to several months. The portfolio comprises between 15 and 30 names at a time. The net asset value (NAV) of this portfolio is shown in Figure 3.7, where an initial value of $1,000$ is set. The returns are net of all fees, including trading costs, management fees, and performance fees and do not include any cash management. In Figure 3.7, the NAV of the BSE200 Index is also reported. BSE200 Index is an index composed by the 200 biggest market capitalizations quoted in the Indian equity market.

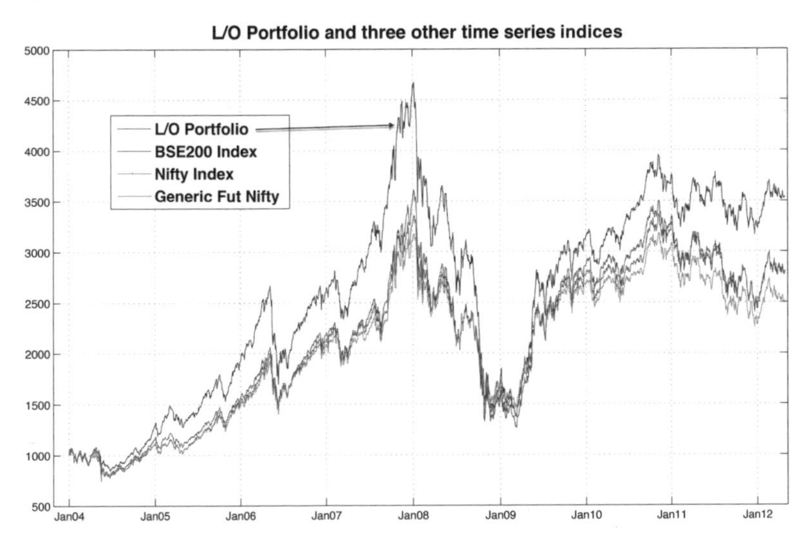

Figure 3.7. *Long-only Indian equity portfolio price and two reference stock indices, the BSE200 Index and the Nifty Index*

The Indian equity stock market is very specific: even if the BSE200 Index is quoted on a daily basis, there are no future contracts quoted for this index. Only future contracts on the smallest Nifty Index are traded, so that even if the BSE200 Index is chosen as the benchmark reference for the performance of the portfolio, it will not be possible to hedge the portfolio by shorting future contracts on the BSE200 Index. The single-factor f is therefore the Nifty Index and the time series used to compute the model parameters is the time series of the generic Nifty Index futures, which is a roll-adjusted time series.

The hedging procedure is presented in the following:

– At each computing time t, parameters α_t and β_t are estimated with method M given, if necessary, a fixed window length.

– The value of the estimated β_t is then used as the parameter that reflects the linear dependence between the past returns of the portfolio and the past returns of the factor: we will then adjust the previous quantity $q_{t-1,t}$ of futures contracts to short, given the required proportion of the portfolio to be hedged at t for the period $[t, t+1]$. The quantity of futures contracts to short at t for the time range $[t, t+1]$ is then given by $q_{t,t+1} = Round\left(NAV_t\, \hat{\beta}_t / CV_t\right)$ and the adjustment of the hedge quantity is given by $q_{t,t+1} - q_{t-1,t}$. In this equation, $Round(.)$ rounds off the value to the lowest integer value, NAV_t denotes the NAV of the portfolio at time t and CV_t denotes the value of the futures contracts at time t, equal to its price time the value of one point of this contract, and $\hat{\beta}_t$ is the estimated value of the β at time t.

Depending on the method used to estimate the value of β, the resulting hedged portfolio will have different behavior, given the fact that the estimation might be lagged, dynamic or jittered with residual outliers.

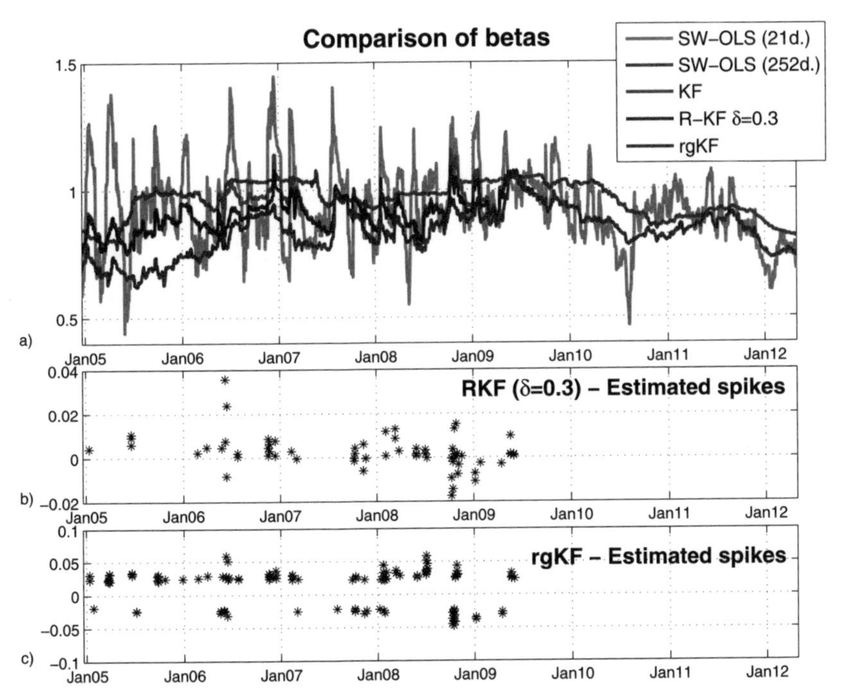

Figure 3.8. *Comparison of the estimated exposure of the L / O portfolio to the selected market factor, the Nifty future: a) represents the dynamic estimates of the beta and b) and c) present the estimated spikes if using an RKF or an rgKG-l^1 as presented in Chapter 4*

Figure 3.8 shows the different results obtained for the estimations of the beta using either a SW-OLS (21 and 252 days), a KF, a Robust Kalman Filter (RKF) with $\delta = 0.3$ and a regularized version of the KF (rgKF-l^1) with $\delta_t = 0.01$ (see Chapter 4 for a detailed description of both these algorithms). Estimating the exposure of the portfolio with a short window length and the OLS yields to very noisy estimates. The longer window gives apparently lagged and too smoothed values. The reference will therefore be the KF results which are more reactive to the beta changes than the SW-OLS and much less noisy. If we compare KF results with the results from more advanced techniques which filter the potential specific events, say first RKF and then rgKF-l^1, combined with the final hedged portfolios and the resulting cumulative costs (see Figure 3.9), then rgKF-l^1 seems to be more efficient and gives more convenient and reliable exposures of the portfolio to the market factor.

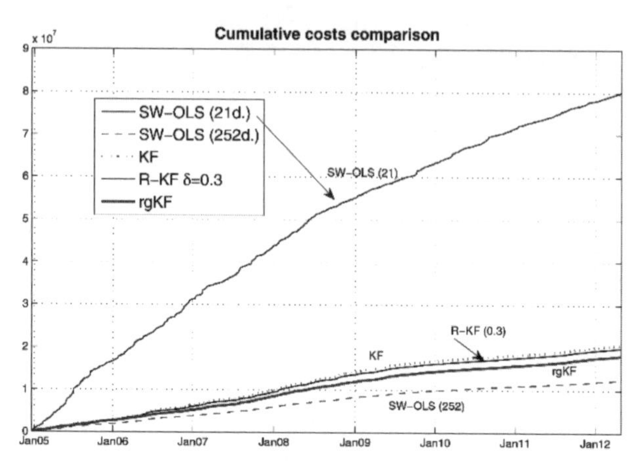

Figure 3.9. *Cumulative costs induced by shorting nifty Index Futures to hedge the L / O portfolio according to each of the estimation method used*

The results are shown in Figure 3.10 and compare the NAVs of the long-only Indian equity portfolio hedged by a proportion of Nifty Index futures whose values are determined either by a constant-beta OLS or a 63 days SW-OLS, or a KF, or a RKF, or a rgKF – see Chapter 4 for a detailed presentation.

From a practical view, the KFs are preferred because it leads to dynamic estimation of the parameters where, at time t, the only knowledge of the past is fully summarized in the previous set of information given in $t - 1$ without any specification of a window length. Moreover, the equations of the KF introduce a dynamic evolution of the model parameters, which brings some *a priori* information about what we actually do not know.

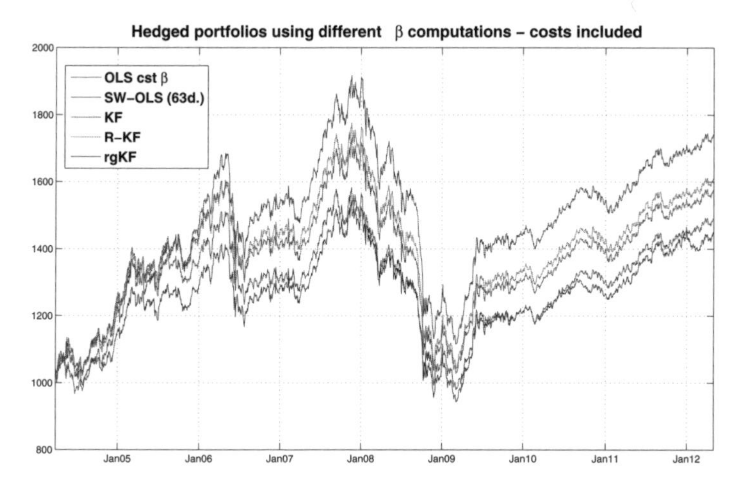

Figure 3.10. *NAVs of a long-only Indian equity portfolio hedged by a proportion of nifty Index futures computed with (1) OLS over the whole period, (2) SW-OLS with a window length of 3 months, (3) a Kalman Filter, (4) a Robust Kalman Filter and (5) a regularized version of the Kalman filter, which filters out outliers. The classical KF and its various versions seem to be the more accurate versus the cost of hedging a portfolio*

Figure 3.10 shows the results for a RKF and a rgKF. KF equations assume conditionally Gaussian-distributed state and observation errors that are invalidated in the presence of outliers. These versions of KF account for outliers in the observation equation and the KF recursive equations are then regularized and therefore robust to outliers. These methods are presented in Chapter 4.

3.13.3. *Hedge fund style analysis using a multi-factor model*

Analyzing the risks in the hedge fund (HF) industry has gained interest because it has been recognized that HF returns have very specific characteristics, such as heavy tail and asymmetric distributions, option-like payoffs and dynamic risk exposures.

Using multi-factor models allows us to decompose the observed returns on a set of several variables, selected as being nonlinearly dependent and bringing multiple insights. The rough task of selecting appropriate factors has been discussed in Chapter 2. In that example, we focused on the estimation methods and use the eight Fung and Hsieh risk factors [HSI].

3.13.3.1. *Fung and Hsieh risk factors*

The eight risk factors of Fung and Hsieh [HSI], as described in Chapter 2, were quoted on a monthly basis from January 1994 to December 2012 and include:

– three *trend-following* risk factors: *PTFSBD*, a bond-based risk factor; *PTFSFX*, a currency-based risk factor, and *PTFSCOM*, a commodity-based risk factor;

– two *equity-oriented* risk factors: *SP500*, the Standard & Poors 500 index, and *EquitySizeSpread*, the size spread factor expressed as the difference between the Russell 2000 index monthly total returns and the *SP500* returns;

– two *bond-oriented* risk factors: *bond10y*, the monthly change in the 10-year treasury constant maturity yield, and *CreditSpread*, the difference between the monthly change in the Moody's Baa yield and *bond10y*;

– one *emerging market* risk factor: *MXEF*, the Morgan Stanley Capital International (MSCI) emerging market index.

These factors are stored in f_t at each time t as inputs for [1.4].

3.13.3.2. *Hedge fund data*

The eight risk factors of Fung and Hsieh were initially built to explain trend-follower funds, so we select *directional* HFs such as *global macro*-hedge funds. Global macro-hedge fund managers may take directional positions in currencies, debt, equities or commodities, and may also elect to take *relative value* positions combining *long* positions paired off against *short* positions.

We illustrate the performance of the different estimation methods on a trend-follower hedge fund, named Aequam Diversified Fund, which has been managed by Aequam Capital since 17 December 2010. The fund is a systematic HF taking long and short positions on futures on bonds, futures on stock indices, futures on foreign exchange and futures on commodities. The fund has been a UCITS IV Fund since June 2012 and is a weekly audited fund with around M$20 of assets under management (AUM). Even though this fund was opened in 2010 we have access to its historical and backtested performance since 2004. Systematic funds have this specificity: backtested performance is reliable because Commodity Trading Advisor (CTA) and systematic funds run a designed process which is exactly the same as the process which is run once the fund is invested. Moreover, managing futures offers a high liquidity to the fund.

Table 3.2 gives some basic statistics for the eight risk factors and for Aequam Diversified Fund. These statistics are computed using monthly (end month-to-end month) returns. Figure 3.11 illustrates the evolution of the eight risk factors of Fung and Hsieh and Aequam Diversified Fund given a base 100 in December 2003.

3.13.3.3. *Estimation methods results*

We compare in this section the results of the estimation of the exposures of the Aequam Diversified Fund for the SW-OLS with 24 months, the KF and the rgKF.

Figure 3.12 displays the tracking performance for Aequam Diversified fund using Fung and Hsieh factors. They have been selected to keep the economic interpretation of the HF exposures. The top graph compares the HF's NAV to the reconstructed NAVs after estimating its returns using OLS, KF and a rgKF. $NAV_1 = 1,000$ at the end of November 2005;

the first 24 months are discarded in regard to the window length used for OLS estimation.

Name	Ann. returns (%)	Ann. volatility (%)	Sharpe ratio	Minimum return (%)	Maximum return (%)	Skewness	Kurtosis	Aequam Diversified Correlation (%)
PTFSBD	−48.55	50.30	−1.01	−26.63	50.50	1.55	5.87	26.85
PTFSFX	−14.84	68.78	−0.24	−30.00	69.22	1.33	4.69	39.33
PTFSCOM	−1.94	50.42	−0.08	−24.65	40.59	0.72	3.15	35.66
SP500	3.87	15.11	0.12	−16.94	10.77	−0.77	4.88	−10.1
EquitySizeSpread	2.60	8.51	0.07	−4.89	6.36	0.25	2.74	−0.26
Bond10y	−6.52	29.16	−0.29	−26.93	27.56	−0.02	4.64	−7.36
CreditSpread	6.47	28.78	0.16	−15.55	38.25	1.24	6.76	23.65
MXEF	12.42	24.87	0.42	−27.50	16.66	−0.69	4.57	2.61
Aequam Diversified	14.11	17.78	0.68	−12.63	16.55	0.50	3.68	–

Table 3.2. *Averaged statistics and correlation for the eight risk factors of Fung and Hsieh and Aequam Diversified Fund. Computation of the statistics is made on monthly returns from January 2004 to December 2012*

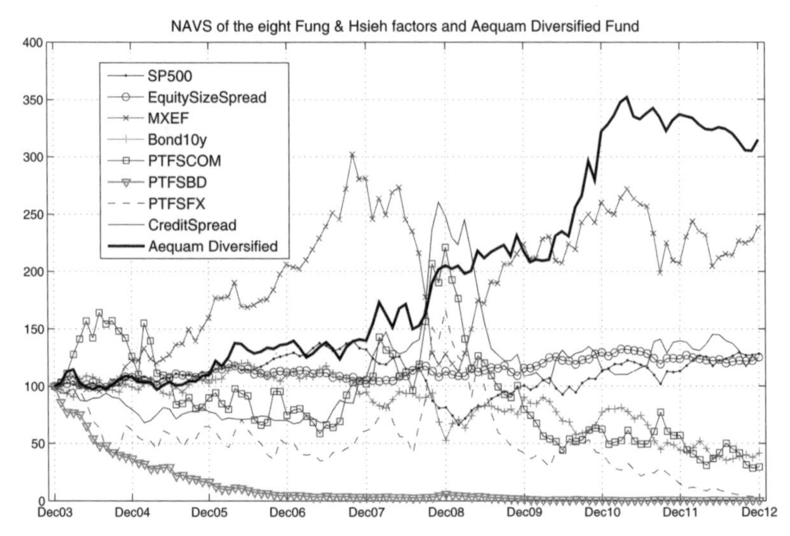

Figure 3.11. *Time series evolution of the eight risk factors of Fung and Hsieh and Aequam Diversified Fund given a base 100 in December 2003 and compounding the monthly returns*

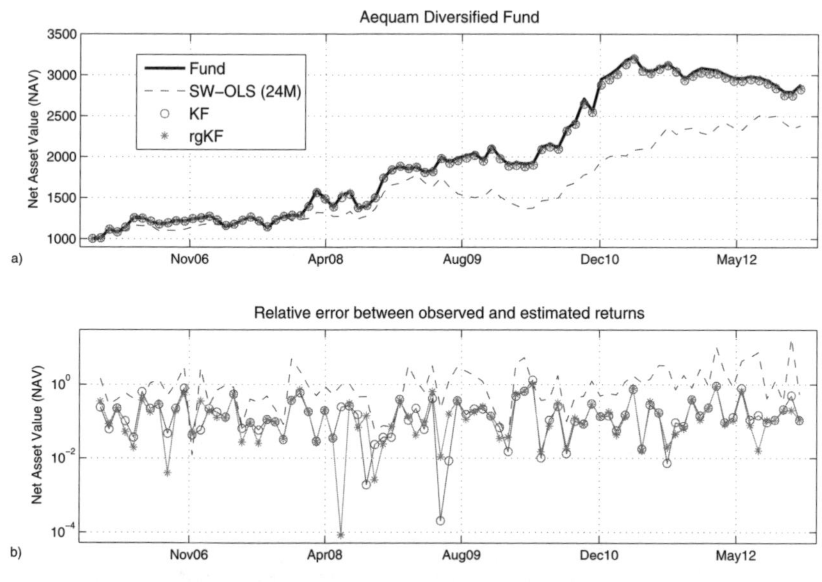

Figure 3.12. *Estimated and observed hedge fund returns for Aequam Diversified Fund (see color plate section)*

Figure 3.12(b) represents the evolution, in a logarithmic scale, of the absolute relative error ($|(r_{i,t} - \hat{r}_{i,t})/r_{i,t}|$) arising from each of the three estimation methods. KF and rgKF achieve almost the same performance whereas SW-OLS is a long way from the observed time series.

Figures 3.13–3.15 show the estimated time-varying exposures for Aequam Diversified Fund using respectively a SW-OLS with a 24 month-long window, a KF and a rgKF. SW-OLS and the KF results are similar but SW-OLS results are more volatile and indicate higher exposures to the factors than KF. The management style of this HF is dynamical but the changes in the estimated exposures are rather smoothed month end to month end. This is exactly what occurs during the management of the fund. The positions are calculated three times a day and adjusted day after day according to a smooth indicator mixing a trend filter and a risk-aversion indicator. Moreover, at the end of the period, Aequam Diversified Fund was long on equity index futures, slightly short on interest rates futures (which is equivalent to be long interest rates), short on US Dollar (USD) versus the major foreign exchanges and short on commodity futures, which is equivalent to being long on the synthetic portfolio of straddles in commodity (PTFSCOM).

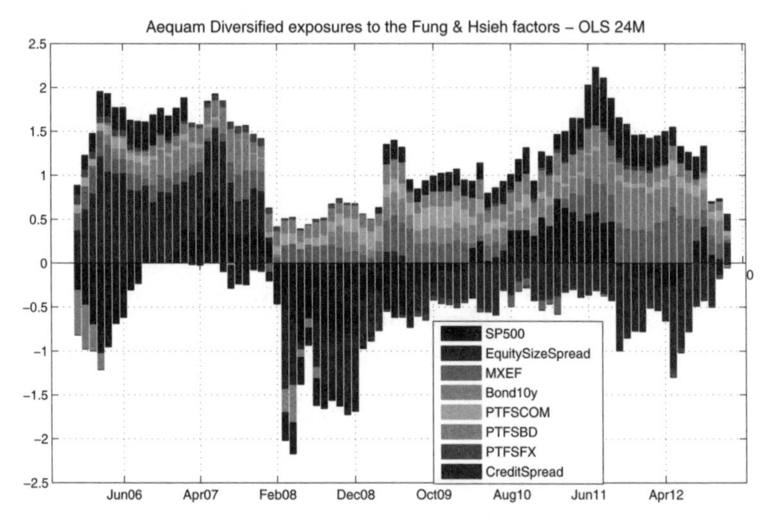

Figure 3.13. *Estimated SW-OLS exposures to the eight risk factors of Fung and Hsieh for the Aequam Diversified Fund (see color plate section)*

The results obtained through this example are coherent with the management style of the fund: dynamic but smoothed exposures and identical exposures versus the real positions. As a preliminary result to the next chapter, it seems that Aequam Diversified Fund does not present any liquidity issues or extreme and specific events because the results of KF and rgKF coincide (no outliers were detected).

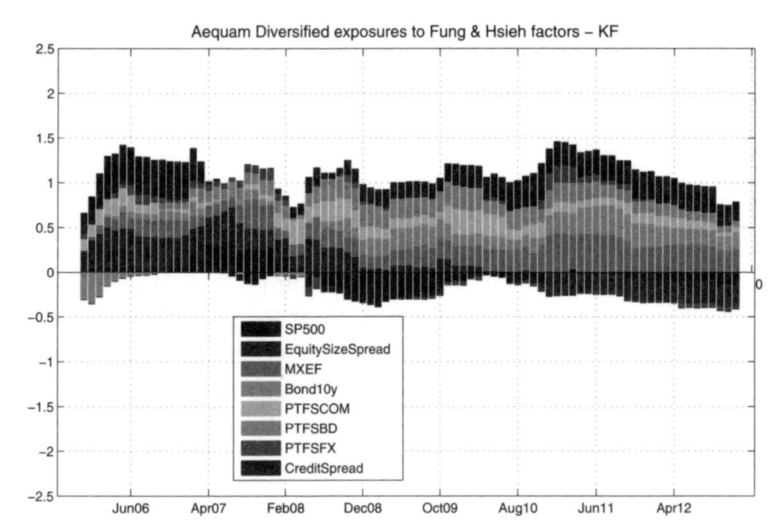

Figure 3.14. *Estimated KF exposures to the eight risk factors of Fung and Hsieh for the Aequam Diversified Fund (see color plate section)*

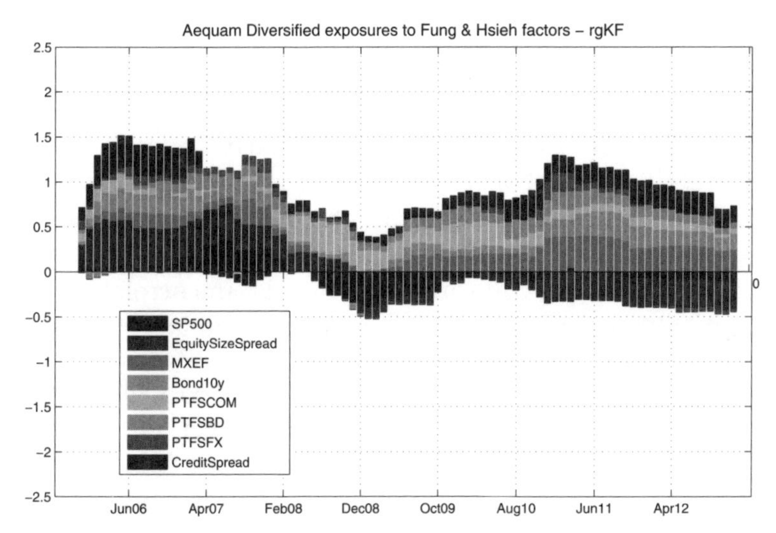

Figure 3.15. *Estimated rgKF exposures to the eight risk factors of Fung and Hsieh for the Aequam Diversified Fund (see color plate section)*

3.14. Geometrical derivation of KF updating equations

3.14.1. *Geometrical interpretation of MSE criterion and the MMSE solution*

KF presentations and derivations are numerous, both in papers and books. In this introductory chapter, instead of following a probabilistic or Bayesian approach that relies on conditional probability densities that often discourage the newcomer who wants to understand where the KF updates come from, we lay a full geometrical foundation of KF, closer to intuitive and Euclidian knowledge, widely shared by most grad readers.

To do so, the first step consists of giving a geometrical interpretation of the problem statement [3.32] and the MMSE criterion at time t – 1. Indeed, KF updates given in section 3.11 rely on the previous estimate at time $t - 1$.

In Figure 3.16, the plan symbolizes the observation space R_1^{t-1} spanned by the random variables $\{r_1, \cdots, r_{t-1}\}$. Any vector element $\boldsymbol{\xi}_{|t-1}$ of R_1^{t-1} is linearly related to the $\{r_k, 1 \leq k \leq t-1\}$ and is represented by a dotted arrow that belongs to the plan. In general, the state $\boldsymbol{\theta}_{t-1}$ at time t – 1 does not belong to the observation plan and thus contains a piece of statistical information that is uncorrelated with (statistically orthogonal to) the random variables $\{r_1, \cdots, r_{t-1}\}$.

The MSE, $\mathbb{E}\left[\|\boldsymbol{\theta}_{t-1} - \boldsymbol{\xi}_{|t-1}\|^2\right]$, can be viewed geometrically as the square of a special Euclidian norm $\langle.|.\rangle_{L^2}$ of the "difference vector" $\boldsymbol{\theta}_{t-1} - \xi_{|t-1}$ drawn in dotted line in Figure 3.16, that is $\mathbb{E}\left[\|\boldsymbol{\theta}_{t-1} - \boldsymbol{\xi}_{|t-1}\|^2\right] = \langle \boldsymbol{\theta}_{t-1} - \xi_{|t-1} | \boldsymbol{\theta}_{t-1} - \xi_{|t-1}\rangle_{L^2}$. As such, the minimum MSE is associated with the minimum geometrical square norm of $\boldsymbol{\theta}_{t-1} - \xi_{|t-1}$ achieved when the element $\xi_{|t-1}$ coincides with the orthogonal projection of $\boldsymbol{\theta}_{t-1}$ onto the observation plan R_1^{t-1}, as depicted in Figure 3.16:

$$\boldsymbol{\xi}_{|t-1}^{op} = \text{Proj}_{\perp}[\boldsymbol{\theta}_{t-1}/R_1^{t-1}] = \hat{\boldsymbol{\theta}}_{t-1|t-1}. \qquad [3.53]$$

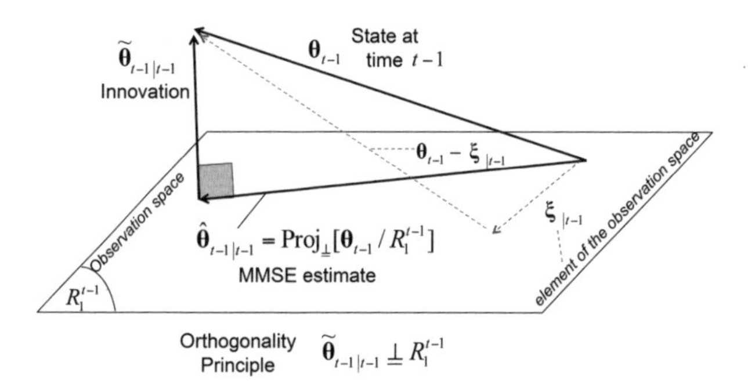

Figure 3.16. *Geometrical interpretation of MMSE estimate*

Orthogonal projection means that the difference vector $\theta_{t-1} - \hat{\theta}_{t-1|t-1}$ that achieves the minimun square norm, called in what follows the innovation, $\tilde{\theta}_{t-1|t-1} = \theta_{t-1} - \hat{\theta}_{t-1|t-1}$ is orthogonal to the observation plan. Being orthogonal to the observation plan implies being orthogonal or, in keeping with the random nature of the variables we deal with, being uncorrelated with all the random variables $\{r_1, \cdots, r_{t-1}\}$.

As a summary, the geometrical light cast on the MMSE, substitutes to the optimization problem a mere orthogonal projection onto the appropriate space (see [3.53]). This can also be expressed in the so-called orthogonality principle: "the innovation is statistically orthogonal to the observation space":

$$\tilde{\theta}_{t-1|t-1} \stackrel{\perp}{=} R_1^{t-1}. \qquad [3.54]$$

The interpretation of the MMSE criterion depicted in Figure 3.16 and more specifically contained in equations [3.53] and [3.54] are the keystones of geometrical derivations of KF updates.

3.14.2. *Derivation of the prediction filtering update*

We derive in this section equation [3.34]. Across prediction filtering, the measurement set is still R_1^{t-1} (no new measurement has been captured w.r.t. the previous estimation). Combination of the recursive component [3.26] of the state-space model, equation [3.54] and the linearity of the projection operator lead to the following identity chain:

$$\begin{aligned}\hat{\boldsymbol{\theta}}_{t|t-1} &= \mathbf{Proj}_{\perp}[\boldsymbol{\theta}_t/R_1^{t-1}] = \mathbf{Proj}_{\perp}[\boldsymbol{\theta}_{t-1} + \mathbf{w}_t/R_1^{t-1}] \\ &= \mathbf{Proj}_{\perp}[\boldsymbol{\theta}_{t-1}/R_1^{t-1}] + \mathbf{Proj}_{\perp}[\mathbf{w}_t/R_1^{t-1}]. \end{aligned} \quad [3.55]$$

Innovation property [3.30] and the statistical orthogonality between $\mathbf{w}_t$ and the observation noise ϵ_t yield $\mathbf{w}_t \perp R_1^{t-1}$ that implies $\mathrm{Proj}_{\perp}[\mathbf{w}_t/R_1^{t-1}] = 0$, which completes the proof of [3.34] (see Figure 3.17).

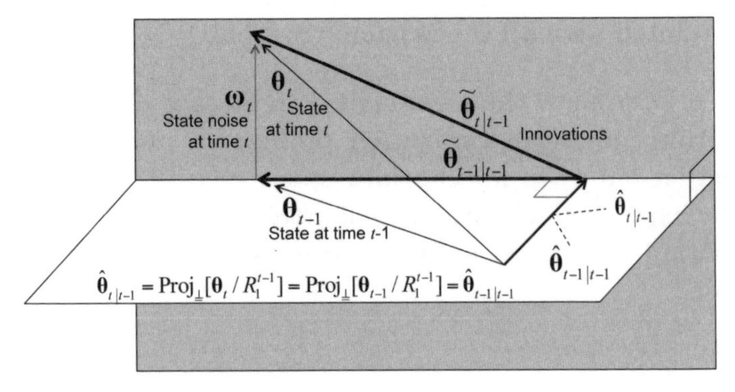

Figure 3.17. *Illustration of the prediction filtering update, see also [3.51] and [3.34]*

3.14.3. *Derivation of the prediction accuracy update*

We derive in this section equation [3.35]. Across the prediction accuracy updating, the measurement set is still R_1^{t-1} (no new measurement has been captured w.r.t. the previous estimation). The definition of the $KF(n/q)$ innovation $\tilde{\boldsymbol{\theta}}_{n|q} = \boldsymbol{\theta}_n - \hat{\boldsymbol{\theta}}_{n|q}$ combined with [3.34] and [3.26] leads to the recursive computation of the "prediction step innovation" $\tilde{\boldsymbol{\theta}}_{t|t-1}$:

$$\tilde{\boldsymbol{\theta}}_{t|t-1} = \tilde{\boldsymbol{\theta}}_{t-1|t-1} + \mathbf{w}_t. \qquad [3.56]$$

Moreover, $\mathbf{w}_t \perp \tilde{\boldsymbol{\theta}}_{t-1|t-1}$ because of the innovation property [3.30] and the following orthogonality $\mathbf{w}_t \perp R_1^{t-1}$ unveiled in section 3.14.2 (see Figure 3.17). Taking into account both the definition of the covariance matrix $\mathbf{S}_{n|q} = \mathbb{E}[\tilde{\boldsymbol{\theta}}_{n|q}\tilde{\boldsymbol{\theta}}'_{n|q}]$ and equation [3.56] added to the fact that the covariance matrix of the summation of two uncorrelated vectors is the summation of the two associated covariance matrices, this completes the proof of [3.35].

3.14.4. *Derivation of the correction filtering update*

We derive in this section both equations [3.36] and [3.37].

3.14.4.1. *Updating of the measurement set*

Correction filtering leverages the new measurement r_t thus involving orthogonal projections (as per section 3.14.1) onto the enlarged (updated) observation (return) set $R_1^t = R_1^{t-1} \cup \{r_t\}$. Because r_t is not necessarily orthogonal to the previous measurement set R_1^{t-1}, the orthogonal projection onto R_1^t is not necessarily equal to the summation of the projection onto R_1^{t-1} and the projection on r_t, as it would be the case, if the projection was operating with an orthogonalized enlarged measurement (return) set. Nevertheless, the correction filtering update requires the expression of $\text{Proj}[./R_1^t]$ in terms of $\text{Proj}[./R_1^{t-1}]$. This becomes possible by "orthogonalizing" the enlarged observation (return) set R_1^t.

3.14.4.2. *Orthogonal decomposition of the updated observation set* $R_1^t = R_1^{t-1} \cup \{r_t\}$

Orthogonalizing the enlarged set R_1^t consists of breaking the new measurement r_t into a component that belongs to R_1^{t-1} plus an orthogonal residual. This orthogonal splitting is straightforwardly derived from the geometrical interpretation described in section 3.14.1. Indeed, applying the orthogonal break of $\boldsymbol{\theta}_{t-1}$ displayed in Figure 3.16 to r_t yields:

$$r_t = \tilde{r}_{t|t-1} + \mathbf{Proj}[r_t/R_1^{t-1}] = \tilde{r}_{t|t-1} + \hat{r}_{t|t-1}, \qquad [3.57]$$

where, by construction, $\tilde{r}_{t|t-1} \perp R_1^{t-1}$. Moreover, the observation component [3.27] of the state-space model allows us to write:

$$\mathbf{Proj}[r_t/R_1^{t-1}] = \mathbf{Proj}[\mathbf{g}_t' \boldsymbol{\theta}_t + \epsilon_t/R_1^{t-1}]. \qquad [3.58]$$

Linearity properties of the projection operator, the uncorrelation (orthogonality) between ϵ_t and R_1^{t-1}, combined with equations [3.27] and [3.57], give the following identities:

$$\tilde{r}_{t|t-1} = r_t - \mathbf{Proj}[r_t/R_1^{t-1}] = r_t - \mathbf{g}_t' \hat{\boldsymbol{\theta}}_{t|t-1} = \mathbf{g}_t' \tilde{\boldsymbol{\theta}}_{t|t-1} + \epsilon_t. \qquad [3.59]$$

Finally, we can write that the updated measurement set R_1^t is split between R_1^{t-1} and the new measurement innovation $\tilde{r}_{t|t-1}$ that happens to be orthogonal to R_1^{t-1} by construction:

$$R_1^t = R_1^{t-1} \oplus \{\tilde{r}_{t|t-1}\} \text{ with } \tilde{r}_{t|t-1} \perp R_1^{t-1}. \qquad [3.60]$$

In [3.60], the symbol $\oplus$ stands for the orthogonal decomposition.

3.14.4.3. *Correction of the predicted estimate* $\hat{\boldsymbol{\theta}}_{t|t-1}$ *based on the new measurement* r_t

In terms of orthogonal projection, the identity of [3.60] helps to split the projection of $\boldsymbol{\theta}_t$ into the enlarged measurement (return) set R_1^t such that:

$$\text{Proj}[\boldsymbol{\theta}_t / R_1^t] = \text{Proj}[\boldsymbol{\theta}_t / R_1^{t-1}] + \text{Proj}[\boldsymbol{\theta}_t / \tilde{r}_{t|t-1}]. \qquad [3.61]$$

In keeping with the geometrical interpretation depicted in Figure 3.16 of the different estimates, equation [3.61] can be rewritten in order to express the new estimate $\hat{\boldsymbol{\theta}}_{t|t}$ as the "predicted estimate" $\hat{\boldsymbol{\theta}}_{t|t-1}$ plus a "correction term" due to the new observation r_t:

$$\hat{\boldsymbol{\theta}}_{t|t} = \hat{\boldsymbol{\theta}}_{t|t-1} + \text{Proj}[\boldsymbol{\theta}_t / \tilde{r}_{t|t-1}]. \qquad [3.62]$$

The correction term is equal to $\text{Proj}[\boldsymbol{\theta}_t / \tilde{r}_{t|t-1}]$ and it has to be explicitly evaluated to obtain [3.36] and [3.37]. Figure 3.18 geometrically illustrates identity [3.62].

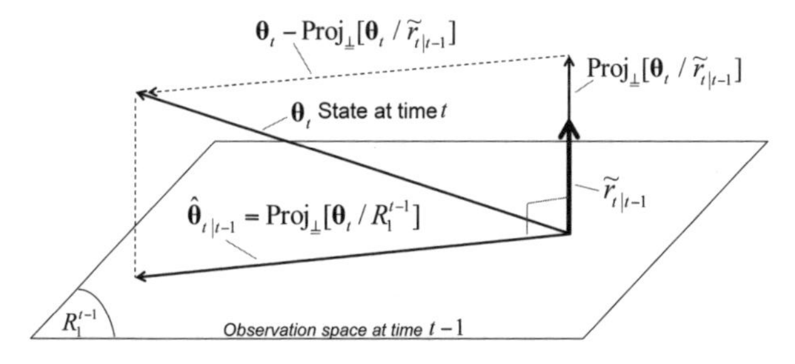

Figure 3.18. *Geometrical illustration of the correction update undergone by the predicted estimate* $\hat{\boldsymbol{\theta}}_{t|t-1}$ *due to the new measurement* r_t*, see also equations [3.59] and [3.62]*

3.14.4.4. *Evaluation of the correction component* $Proj_\perp[\boldsymbol{\theta}_t / \tilde{r}_{t|t-1}]$

Like any projection and as illustrated in Figure 3.18, the orthogonal projection of $\boldsymbol{\theta}_t$ on the measurement innovation

$\tilde{r}_{t|t-1}$ is "proportional" to $\tilde{r}_{t|t-1}$. But since $\boldsymbol{\theta}_t$ is a vector and $\tilde{r}_{t|t-1}$ a scalar, the proportionality relationship involves a vector $\mathbf{z}_t$ such that:

$$\mathbf{Proj}_{\perp}[\boldsymbol{\theta}_t/\tilde{r}_{t|t-1}] = \mathbf{z}_t\,\tilde{r}_{t|t-1}. \qquad [3.63]$$

Noting that, by construction and in keeping with Figure 3.18, the "difference" $\boldsymbol{\theta}_t - \mathbf{Proj}_{\perp}[\boldsymbol{\theta}_t/\tilde{r}_{t|t-1}]$ is statistically orthogonal to $\tilde{r}_{t|t-1}$ which allows us to derive the explicit form of $\mathbf{z}_t$. Indeed, the definition of statistical orthogonality, combined with [3.63], yields the following chain:

$$\boldsymbol{\theta}_t - \mathbf{Proj}_{\perp}[\boldsymbol{\theta}_t/\tilde{r}_{t|t-1}] \perp \tilde{r}_{t|t-1} \Rightarrow \mathbb{E}\left[(\boldsymbol{\theta}_t - \mathbf{z}_t\,\tilde{r}_{t|t-1})\,\tilde{r}_{t|t-1}\right] = 0. \qquad [3.64]$$

Chain [3.64] straightfowardly leads to:

$$\mathbf{z}_t = \frac{\mathbb{E}[\boldsymbol{\theta}_t\,\tilde{r}_{t|t-1}]}{\mathbb{E}[\tilde{r}^2_{t|t-1}]}. \qquad [3.65]$$

Taking into account that $\mathbb{E}[\tilde{\boldsymbol{\theta}}_{t|t-1}\,\tilde{\boldsymbol{\theta}}'_{t|t-1}] = \mathbf{S}_{t|t-1}$ and equation [3.59] added to the orthogonality between ϵ_t and $\tilde{\boldsymbol{\theta}}_{t|t-1}$ help us to write the denominator of $\mathbf{z}_t$ as:

$$\mathbb{E}[\tilde{r}^2_{t|t-1}] = \mathbf{g}'_t\,\mathbf{S}_{t|t-1}\,\mathbf{g}_t + \sigma^2_{\epsilon}. \qquad [3.66]$$

According to equation [3.60] and Figure 3.18, we have $\tilde{r}_{t|t-1} \perp R_1^{t-1}$; therefore, $\mathbb{E}[\hat{\boldsymbol{\theta}}_{t|t-1}\,\tilde{r}_{t|t-1}] = 0$, which combined with [3.55], allows us to write the following different forms of the numerator of $\mathbf{z}_t$ [3.36] as:

$$\mathbb{E}[\boldsymbol{\theta}_t\,\tilde{r}_{t|t-1}] = \mathbb{E}[(\boldsymbol{\theta}_t - \hat{\boldsymbol{\theta}}_{t|t-1})\,\tilde{r}_{t|t-1}] = \mathbb{E}[\tilde{\boldsymbol{\theta}}_{t|t-1}\,\tilde{r}_{t|t-1}] = \mathbf{S}_{t|t-1}\,\mathbf{g}_t. \qquad [3.67]$$

Finally equations [3.63] and [3.65]–[3.67] give:

$$\mathbf{Proj}_{\perp}[\boldsymbol{\theta}_t/\tilde{r}_{t|t-1}] = \mathbf{z}_t\,\tilde{r}_{t|t-1} = \frac{\mathbf{S}_{t|t-1}\,\mathbf{g}_t}{\mathbf{g}'_t\,\mathbf{S}_{t|t-1}\,\mathbf{g}_t + \sigma^2_{\epsilon}}\,\tilde{r}_{t|t-1}. \qquad [3.68]$$

3.14.4.5. *Correction filtering updates, equations [3.36] and [3.37]*

We derive from [3.59], [3.62] and [3.68] the following correction filtering update chain due to the capture of the new measurement r_t:

$$\hat{\boldsymbol{\theta}}_{t|t} = \hat{\boldsymbol{\theta}}_{t|t-1} + \mathbf{z}_t\, \tilde{r}_{t|t-1} = \hat{\boldsymbol{\theta}}_{t|t-1} + \mathbf{z}_t\, (r_t - \mathbf{g}_t'\, \hat{\boldsymbol{\theta}}_{t|t-1})$$

$$\text{with } \mathbf{z}_t = \frac{\mathbf{S}_{t|t-1}\, \mathbf{g}_t}{\mathbf{g}_t'\, \mathbf{S}_{t|t-1}\, \mathbf{g}_t + \sigma_\epsilon^2}, \qquad [3.69]$$

which coincides with [3.36] and [3.37] noting that $\mathbf{z}_t$ is nothing but the Kalman gain.

3.14.5. *Derivation of the correction accuracy update*

We derive in this section equation [3.38]. From the first identity of the chain [3.69], we have:

$$\boldsymbol{\theta}_t - \hat{\boldsymbol{\theta}}_{t|t} = \boldsymbol{\theta}_t - \hat{\boldsymbol{\theta}}_{t|t-1} - \mathbf{z}_t\, \tilde{r}_{t|t-1}, \qquad [3.70]$$

which combined with [3.59] and the definition of the innovation $\tilde{\boldsymbol{\theta}}_{t|t-1}$ implies:

$$\tilde{\boldsymbol{\theta}}_{t|t} = (\mathbf{I} - \mathbf{z}_t\, \mathbf{g}_t')\, \tilde{\boldsymbol{\theta}}_{t|t-1} - \mathbf{z}_t\, \epsilon_t. \qquad [3.71]$$

Moreover, because of the orthogonality principle, that is $\tilde{\boldsymbol{\theta}}_{t|t} \perp \hat{\boldsymbol{\theta}}_{t|t}$, the following identity holds:

$$\mathbf{S}_{t|t} = \mathbb{E}[\tilde{\boldsymbol{\theta}}_{t|t}\, \boldsymbol{\theta}_t']. \qquad [3.72]$$

Substituting [3.71] into [3.72] and the fact that $\boldsymbol{\theta}_t \perp \epsilon_t$ yield:

$$\mathbf{S}_{t|t} = (\mathbf{I} - \mathbf{z}_t\, \mathbf{g}_t')\, \mathbb{E}[\tilde{\boldsymbol{\theta}}_{t|t-1}\, \boldsymbol{\theta}_t']. \qquad [3.73]$$

Finally the identity is:

$$\mathbf{S}_{t|t-1} = \mathbb{E}[\tilde{\boldsymbol{\theta}}_{t|t-1}\,\boldsymbol{\theta}'_t], \qquad [3.74]$$

due also to the orthogonality principle (i.e. $\tilde{\boldsymbol{\theta}}_{t|t-1} \perp \hat{\boldsymbol{\theta}}_{t|t-1}$), see Figure 3.17, transforming equation [3.38] into [3.73] which completes its proof.

3.15. Highlights

This section gives a summary of the most important outcomes of the chapter dealing with LSE and KF.

Least squares estimation (LSE)

LSE implicitly assumes that a linear regression holds between the returns, the factors, the alpha and the betas, across a certain observation window, $(t, h) = \{l,$ such that: $t - h + 1 \leq l \leq t\}$; this regression is also called the LSE "block" measurement equation, it is as follows:

$$\mathbf{r}_{(t,h)} = \mathbf{G}_{(t,h)}\,\boldsymbol{\theta}_{(t,h)} + \boldsymbol{\epsilon}_{(t,h)}. \qquad [3.75]$$

The inputs of the LSE block regression are: the values of one specific return within the window $r'_{(t,h)} = \left[r_t\; r_{t-1} \cdots r_{t-h+1}\right]$ and the factors contained in the $[h, K + 1]$ "factor matrix", $\mathbf{G}_{(t,h)} = rows(\mathbf{g}'_l)$, $t - h + 1 \leq l \leq t$, with $\mathbf{g}'_l = \left[1\; f_{1,l}\; f_{2,l} \cdots f_{K,l}\right]$. Factors are also named the regressors. The unknown or the parameters of the LSE regression are the alpha and betas that are assumed constant across the window. They filled up the column-vector $\theta_{(t,h)}$ with $\boldsymbol{\theta}'_{(t,h)} = \left[\alpha_{(t,h)}\; b_{(t,h),1}\; b_{(t,h),2} \cdots b_{(t,h),K}\right]$. The residual of the LSE regression is $\boldsymbol{\epsilon}_{(t,h)}$ with $\boldsymbol{\epsilon}'_{(t,h)} = \left[\epsilon_t\; \epsilon_{t-1} \cdots \epsilon_{t-h+1}\right]$.

The LSE $\hat{\boldsymbol{\theta}}_{(t,h)}$ minimizes the (convex) square Euclidian norm of the difference between the measurement (return) vector $\mathbf{r}_{(t,h)}$ and its linear approximation based on the factors, that is $\mathbf{G}_{(t,h)}\,\boldsymbol{\xi}$. Assuming that the $[h, K+1]$ matrix $\mathbf{G}_{(t,h)}$ is full rank, the LSE solution is unique and we have:

$$\hat{\boldsymbol{\theta}}_{(t,h)} = \arg\{ \min_{\boldsymbol{\xi}\in\mathbb{R}^{K+1}} \|\mathbf{r}_{(t,h)} - \mathbf{G}_{(t,h)}\,\boldsymbol{\xi}\|_{L^2}^2 \}. \qquad [3.76]$$

The LSE $\hat{\boldsymbol{\theta}}_{(t,h)}$ is an unbiased estimate of $\boldsymbol{\theta}_{(t,h)}$ but its variance may become large when the matrix $(\mathbf{G}'\,\mathbf{G})_{(t,h)}$ is ill conditioned. This happens if its minimum eigenvalue becomes too small. In this latter situation, the LSE has to be "regularized" to limit the magnitude of $\hat{\boldsymbol{\theta}}_{(t,h)}$ (this topic is addressed in Chapter 4). The term $(\mathbf{G}\,\hat{\boldsymbol{\theta}})_{(t,h)}$ is the orthogonal projection of the return vector $\mathbf{r}_{(t,h)}$ into the space spanned by the columns of the factor matrix $\mathbf{G}_{(t,h)}$. The LSE $\hat{\boldsymbol{\theta}}_{(t,h)}$ operates in a block processing fashion and is not recursive. It has thus limited tracking capabilities.

Kalman setup, objective and geometrical view

The KF setup bears two enabling features that naturally lead to a recursive estimate $\hat{\boldsymbol{\theta}}_{t|t}$ of the parameter vector $\boldsymbol{\theta}_t$ (whose components are the alpha and betas). These two key characteristics are missing in the LSE model. First and as opposed to LSE, KF set up leverages an "instantaneous" measurement equation (involving the return, the factors, the alpha and the betas). KF does not need any windowing of the data, as LSE does: LSE equation [3.75] is a block and not instantaneous measurement equation. Indeed, KF observation (return) equation takes the following "instantaneous" form:

$$r_t = \mathbf{g}_t'\,\boldsymbol{\theta}_t + \epsilon_t,\; t \geq 1, \qquad [3.77]$$

with $\mathbf{g}_t' = \left[1\; f_{1,t}\; f_{2,t} \cdots f_{K,t}\right]$ and $\boldsymbol{\theta}_t' = \left[\alpha_t\; b_{t,1}\; b_{t,2} \cdots b_{t,K}\right]$. Second, KF setup relies on a recursive equation that drives the dynamic evolution of $\boldsymbol{\theta}_t$. This is called the state recursive equation ($\boldsymbol{\theta}_t$ is a function of the previous value $\boldsymbol{\theta}_{t-1}$). The recursive equation that does not exist in the LSE model may be derived from the laws of physics, econometrics, finance, etc., or it may be intuitively postulated. In this latter and frequent situation, its form has to be parsimonious. The simplest intuitive "state recursive" equation is as follows:

$$\boldsymbol{\theta}_t = \boldsymbol{\theta}_{t-1} + \mathbf{w}_t,\ t \geq 1, \qquad [3.78]$$

where $\mathbf{w}_t$ is a zero-mean, white, Gaussian noise. The above recursive parameters (i.e. the alpha and betas) equation means that the new "state" of the parameters at time t is equal to the previous "state" at time $t-1$ plus another component, the state noise $\mathbf{w}_t$ that is statistically orthogonal (uncorrelated) to the previous state that is $\mathbf{w}_t \perp \boldsymbol{\theta}_{t-1}$.

Given the recursive evolution equation of the alpha and betas [3.78] and the collection of returns $\{r_k, 1 \leq k \leq t\}$ that are assumed following the observation equation [3.77], KF provides a recursive (and linear) estimate $\hat{\boldsymbol{\theta}}_{t|t}$ that minimizes the MSE $\mathbb{E}\left[\|\boldsymbol{\theta}_t - \boldsymbol{\xi}_{|t}\|_{L^2}\right]$ ($\mathbb{E}[.]$ designates the expected value). The vector $\boldsymbol{\xi}_{|t}$ belongs to the space R_1^t spanned by the returns $\{r_k, 1 \leq k \leq t\}$. As defined, the KF optimality criterion benefits from the same (natural) geometrical interpretation as the LSE Euclidian metrics, provided the MSE is viewed as the square of a special Euclidian norm $\langle .|. \rangle_{L^2}$, that is $\mathbb{E}\left[\|\boldsymbol{\theta}_{t-1} - \boldsymbol{\xi}_{|t-1}\|_{L^2}^2\right] = \langle \boldsymbol{\theta}_{t-1} - \xi_{|t-1} | \boldsymbol{\theta}_{t-1} - \xi_{|t-1} \rangle_{L^2}$.

Kalman filtering solution and updates

KF provides updating not only for the previous estimate $\hat{\boldsymbol{\theta}}_{t-1|t-1}$ but also for the previous covariance or "accuracy" matrix, $\mathbf{S}_{t-1|t-1} = \mathbb{E}\left[(\boldsymbol{\theta}_{t-1} - \hat{\boldsymbol{\theta}}_{t-1|t-1})(\boldsymbol{\theta}_{t-1} - \hat{\boldsymbol{\theta}}_{t-1|t-1})'\right]$, as suggested by the MSE criterion.

The two inputs of the full KF updating step from $t-1|t-1$ to $t|t$ are thus the "previous" estimate and "previous" covariance matrix, that is $\hat{\boldsymbol{\theta}}_{t-1|t-1}$ and $\mathbf{S}_{t-1|t-1}$ with $t \geq 1$. This full KF step is split between the prediction and the correction steps.

– **Prediction step:** from $t-1|t-1$ to $t|t-1$

Inputs: $\hat{\boldsymbol{\theta}}_{t-1|t-1}$, $\mathbf{S}_{t-1|t-1}$ → Output: $\hat{\boldsymbol{\theta}}_{t|t-1}$, $\mathbf{S}_{t|t-1}$ ("predicted" estimate and "predicted" covariance matrix).

The prediction transition leverages the recursive evolution equation [3.78] to "recursively" anticipate the evolution of the estimate and its accuracy based only on the predicted parameter, according to the model [3.78]. The prediction step does not make use of a new measurement (current value of the return). It comprises the prediction filtering and the prediction accuracy processing:

Prediction filtering

$$\hat{\boldsymbol{\theta}}_{t|t-1} = \hat{\boldsymbol{\theta}}_{t-1|t-1} \quad [3.79]$$

Prediction accuracy processing

$$\mathbf{S}_{t|t-1} = \mathbf{S}_{t-1|t-1} + \sigma_w^2 \mathbf{I} \quad [3.80]$$

Because the prediction transition does not benefit from a new value of the return, it loses accuracy as displayed in equation [3.80], that is $tr[\mathbf{S}_{t|t-1}] \geq tr[\mathbf{S}_{t-1|t-1}]$.

– **Correction step:** from $t|t-1$ to $t|t$

Inputs (predicted estimate and covariance matrix): $\hat{\boldsymbol{\theta}}_{t|t-1}$, $\mathbf{S}_{t|t-1}$ → Outputs: $\hat{\boldsymbol{\theta}}_{t|t}$, $\mathbf{S}_{t|t}$ ("current" estimate and "current" covariance matrix).

The correction step leverages the instantaneous observation equation [3.78] and the current value of the return r_t. The correction step comprises the correction filtering and the correction accuracy processing. The correction filtering balances, via the Kalman gain $\mathbf{z}_t$ the weight of the predicted estimate $\hat{\boldsymbol{\theta}}_{t|t-1}$ and the weight of the current measurement r_t:

Correction filtering

$$\hat{\boldsymbol{\theta}}_{t|t} = \hat{\boldsymbol{\theta}}_{t|t-1} + \mathbf{z}_t \left(r_t - \mathbf{g}_t' \, \hat{\boldsymbol{\theta}}_{t|t-1} \right) \quad [3.81]$$

Correction accuracy processing

$$\mathbf{S}_{t|t} = (\mathbf{I} - \mathbf{z}_t \, \mathbf{g}_t') \, \mathbf{S}_{t|t-1} \quad [3.82]$$

The Kalman gain $\mathbf{z}_t = \mathbf{S}_{t|t-1} \, \mathbf{g}_t \left(\mathbf{g}_t' \, \mathbf{S}_{t|t-1} \, \mathbf{g}_t + \sigma_\epsilon^2 \right)^{-1}$ decreases if the current measurement is not reliable (large variance σ_ϵ^2 of the measurement noise) or if the algorithm has converged close to the true parameter (covariance matrix $\mathbf{S}_{t|t-1}$ is "small"). In these two latter situations, priority is given to the predicted estimate $\hat{\boldsymbol{\theta}}_{t|t-1}$. Conversely, the Kalman gain increases if the observation noise becomes marginal, that is $\sigma_\epsilon^2 \to 0$. KF is thus self-regulated via its gain.

3.16. Appendix: Matrix inversion lemma

In this appendix, we prove identity [3.44] based on direct calculation of the following matrix $\mathbf{P}$:

$$\mathbf{P} = \left(\mathbf{S}_{t|t-1} - \frac{\mathbf{S}_{t|t-1} \, \tilde{\mathbf{g}}_t \, \tilde{\mathbf{g}}_t' \, \mathbf{S}_{t|t-1}}{1 + \tilde{\mathbf{g}}_t' \, \mathbf{S}_{t|t-1} \, \tilde{\mathbf{g}}_t} \right) \left(\mathbf{S}_{t|t-1}^{-1} + \tilde{\mathbf{g}}_t \, \tilde{\mathbf{g}}_t' \right). \quad [3.83]$$

Noting that the quadratic form $\tilde{\mathbf{g}}_t' \, \mathbf{S}_{t|t-1} \, \tilde{\mathbf{g}}_t$ is a scalar, we straightly derive from [3.83] that:

$$\mathbf{P} = \mathbf{I} - \frac{\mathbf{S}_{t|t-1} \, \tilde{\mathbf{g}}_t \, \tilde{\mathbf{g}}_t'}{1 + \tilde{\mathbf{g}}_t' \, \mathbf{S}_{t|t-1} \, \tilde{\mathbf{g}}_t} + \mathbf{S}_{t|t-1} \, \tilde{\mathbf{g}}_t \, \tilde{\mathbf{g}}_t' - \frac{\tilde{\mathbf{g}}_t' \, \mathbf{S}_{t|t-1} \, \tilde{\mathbf{g}}_t \, \mathbf{S}_{t|t-1} \, \tilde{\mathbf{g}}_t \, \tilde{\mathbf{g}}_t'}{1 + \tilde{\mathbf{g}}_t' \, \mathbf{S}_{t|t-1} \, \tilde{\mathbf{g}}_t}. \quad [3.84]$$

In [3.84], $\mathbf{I}$ designates the identity matrix. On the right-hand side of [3.84], reduction to the same denominator of $\mathbf{S}_{t|t-1} \, \tilde{\mathbf{g}}_t \, \tilde{\mathbf{g}}_t'$ yields finally the identity $\mathbf{P} = \mathbf{I}$ that completes the proof of [3.44].

Chapter 4

A Regularized Kalman Filter (rgKF) for Spiky Data

4.1. Introduction

This chapter is an extension of the previous chapter and presents a regularized version of the Kalman filter (KF). As stated earlier, the recursive equations of the KF rely on the Gaussian assumption of the errors of the model so that exogenous outliers coming from either unknown sensor failures, abnormal measurements, portfolio illiquidity issues or even intentional jamming challenge the state-space model introduced in Chapter 3.

In the framework of tracking or estimating unobserved processes, usually solved with KF or one of its extensions, such exogenous outliers are referred to as *additive outliers* or *AOs* [FOX 72, RUC 10] and occur in the observation. Endogenous outliers that may affect the innovations of the state equation are called *innovative outliers* or *IOs* and are beyond the scope of this book. Recursive least squares can serve to deal with AOs applying a Huber weight function

[MAR 06] to the innovation error in the KF recursive equation. We can also penalize the KF by introducing an l^1- or l^2-regularization term. Regularization usually offers good results but requires the determination of the regularization parameter introduced to limit the amplitude of the solution. The search for the optimal regularization parameter has no ideal solution and each existing method (e.g. cross-validation, generalized cross-validation [CRA 79], maximum *a posteriori* estimation [PAN 11], Bernouilli-Gaussian assumption or Markov Chain Monte Carlo (MCMC) methods [AND 97]) is relevant to the problem under study. Regularization is widely used in statistical signal processing (e.g. in basis pursuit, compressed sensing, signal recovery and wavelet thresholding), in statistics with ridge regressions and Least Absolute Shrinkage and Selection Operator (LASSO) algorithm or fused LASSO [TIB 96] and is also very useful in decoding linear codes or geophysics problems. In finance, l^1-regularization also applies in the context of portfolio optimization [JAG 03, BRO 08, DEM 09].

In this chapter, we first discuss a robust Kalman filter (RKF) as presented in [MAT 10], whose performance depends on the value of the regularization parameter, and then propose an alternative methodology to the choice of the regularization parameter by introducing an outlier *detection* step in the recursive equations of the KF. Detecting first the presence of the outliers is equivalent to implicitly choosing the value of the regularization parameter and remains valid even if there is no outlier.

This chapter is organized as follows: section 4.3 discusses the RKF as described in [MAT 10]. The family of *detection and estimation* algorithms that we introduce to handle exogenous outliers is discussed in section 4.4. Section 4.5 deals with the application of our algorithm to detect irregularities in hedge fund returns.

4.2. Preamble: statistical evidence on the KF recursive equations

If we have a look to the recursive equations of the KF (summarized in the table below), we clearly realize that the observation r_t, the unknown and unobserved states $\boldsymbol{\theta}_t$ and the innovation are Gaussian processes. If a large error measurement (most often considered an outlier) occurs at t, the observation r_t is then corrupted as well as the innovation term. It will yield to an erroneous correction for $\hat{\boldsymbol{\theta}}_{t/t}$ whereas the accuracy matrix $\mathrm{S}_{t/t}$ and the Kalman gain $\mathbf{z}_t$ will not be impacted. The error will spread along the recursions especially if the error variance is fixed.

	Equation	Hypothesis	Distribution
State Observation	$\boldsymbol{\theta}_t = \boldsymbol{\theta}_{t-1} + \mathbf{w}_t$ $r_t = \mathbf{g}_t' \boldsymbol{\theta}_t + \epsilon_t$	$\mathbf{w}_t \sim \mathcal{N}_{K+1}(\mathbf{0}, \sigma_w^2 \mathbf{I})$ $\epsilon_t \sim \mathcal{N}(0, \sigma_\epsilon^2)$	$\boldsymbol{\theta}_t \sim \mathcal{N}(\bar{\boldsymbol{\theta}}_{t-1}, \mathbf{S}_{t-1} + \sigma_w^2 \mathbf{I})$ $r_t \sim \mathcal{N}(\mathbf{g}_t' \bar{\boldsymbol{\theta}}_t, \mathbf{g}_t' \mathbf{S}_t \mathbf{g}_t + \sigma_\epsilon^2)$
Prediction	$\hat{\boldsymbol{\theta}}_{t\|t-1} = \hat{\boldsymbol{\theta}}_{t-1\|t-1}$ $\mathbf{S}_{t\|t-1} = \mathbf{S}_{t-1\|t-1} + \sigma_w^2 \mathbf{I}$	$\boldsymbol{\theta}_0 \sim \mathcal{N}_{K+1}(\boldsymbol{\mu}_0, \sigma_0^2 \mathbf{I})$ $\mathbf{S}_{0\|0} = \sigma_0^2 \mathbf{I}$; $(\sigma_0^2, \boldsymbol{\mu}_0)$ known	
Kalman Gain Innovation	$\mathbf{z}_t = \mathbf{S}_{t\|t-1} \mathbf{g}_t (\mathbf{g}_t' \mathbf{S}_{t\|t-1} \mathbf{g}_t + \sigma_\epsilon^2)^{-1}$ $\hat{\epsilon}_{t\|t-1} = r_t - \mathbf{g}_t' \hat{\boldsymbol{\theta}}_{t\|t-1}$		$\hat{\epsilon}_{t\|t-1}$ Gaussian
Correction	$\hat{\boldsymbol{\theta}}_{t\|t} = \hat{\boldsymbol{\theta}}_{t\|t-1} + \mathbf{z}_t \hat{\epsilon}_{t\|t-1}$ $\mathbf{S}_{t\|t} = (\mathbf{I} - \mathbf{z}_t \mathbf{g}_t') \mathbf{S}_{t\|t-1}$		

In such a case, the state-space model may be extended with an additional term in the observation equation, which will be taken into account in the recursive estimates. The following section describes how to deal with such an extended model and discusses the RKF (see [MAT 10]).

4.3. Robust KF

4.3.1. *RKF description*

The setup of the RKF introduces an additional term in [3.27] to capture outliers that may appear in the observations. The extended state-space model is written as follows [MAT 10]:

$$\boldsymbol{\theta}_t = \boldsymbol{\theta}_{t-1} + \boldsymbol{\omega}_t, \qquad [4.1]$$

$$r_t = \mathbf{g}_t' \boldsymbol{\theta}_t + v_t + \epsilon_t. \qquad [4.2]$$

The basic assumptions are the same as for the state-space model [3.27] and v_t is a stochastic impulse process, for example a Bernoulli–Gaussian process, suitable for modeling unknown sensor failures, measurement outliers or even intentional jamming. As with the standard KF (where $v_t = 0$), the objective of the problem stated in [4.1] and [4.2] is to derive recursive estimate $\hat{\boldsymbol{\theta}}_{t|t}$ of the unobservable parameter $\boldsymbol{\theta}_t$, given the measurement r_t and the observed variables $\mathbf{g}_t$ up to time t, based on the minimization of the mean-square error (MSE) $\mathbb{E}(|\boldsymbol{\theta}_t - \hat{\boldsymbol{\theta}}_{t|t}|^2)$.

In the presence of the non-Gaussian outlier v_t, the optimum recursive estimate cannot be explicitly derived and moreover may be nonlinear. An elegant alternative consists of regularizing the KF quadratic cost function by adding a l^q-penalty (l^1 in RKF).

The RKF algorithm consists of following steps:

$$\text{Prediction} \quad \hat{\boldsymbol{\theta}}_{t|t-1} = \hat{\boldsymbol{\theta}}_{t-1|t-1}, \qquad [4.3]$$

$$\mathbf{S}_{t|t-1} = \mathbf{S}_{t-1|t-1} + \sigma_w^2 \mathbf{I}, \qquad [4.4]$$

$$\text{Innovation} \quad \hat{\epsilon}_{t|t-1} = r_t - \mathbf{g}_t' \hat{\boldsymbol{\theta}}_{t|t-1}, \qquad [4.5]$$

$$\text{Kalman gain} \quad \mathbf{z}_t = \mathbf{S}_{t|t-1} \mathbf{g}_t \left(\mathbf{g}_t' \mathbf{S}_{t|t-1} \mathbf{g}_t + \sigma_\epsilon^2\right)^{-1} \qquad [4.6]$$

$$\text{Correction} \quad \mathbf{S}_{t|t} = (\mathbf{I} - \mathbf{z}_t \mathbf{g}_t') \mathbf{S}_{t|t-1}, \qquad [4.7]$$

$$\hat{\boldsymbol{\theta}}_t = \hat{\boldsymbol{\theta}}_{t|t-1} + \mathbf{z}_t \left(\hat{\epsilon}_{t|t-1} - \hat{v}_t\right), \qquad [4.8]$$

Regularization $\hat{v}_t$ is a solution of $\min_{v_t} C(v_t, \delta)$

$$= \min_{v_t} (\hat{\epsilon}_{t|t-1} - v_t)' q_t (\hat{\epsilon}_{t|t-1} - v_t) + \delta \|v_t\|_1 \qquad [4.9]$$

with

$$q_t = (1 - \mathbf{g}_t' \mathbf{z}_t)' \sigma_\epsilon^{-2} (1 - \mathbf{g}_t' \mathbf{z}_t) + \mathbf{z}_t' \mathbf{S}_{t|t-1}^{-1} \mathbf{z}_t, q_t > 0 \qquad [4.10]$$

and $\delta \geq 0$ is the regularization parameter. [4.11]

Equations [4.3]–[4.7] are similar to the KF recursive equations. Equation [4.9] is referred to as a *regularized regression* and is equivalent to the LASSO setup [TIB 96, OSB 00]. Given an appropriate value of δ, minimizing $C(v_t, \delta)$ is equivalent to both detecting and estimating v_t. The l^1-norm will reduce the value of v_t to zero for high values of δ and consider a non-zero value of v_t with a small value of δ (see Figure 4.1 for an illustration). Jointly detecting and estimating $\hat{v}_t$ depends therefore on the good choice of δ, which has to be tuned such that the sparsity of v_t coincides with the sparsity seen through the observations. If the value of δ is inappropriate, then many false alarms are generated. In [TIB 96], the author observes that the search for δ can be restricted to the range $[0, 2\, q_t\, |\hat{\epsilon}_{t|t-1}|]$. In practice, the upper bound seems to be very generous [SCH 05], but it can be adaptive and specific to the observed variable.

4.4. rgKF: the rgKF(NG,l^q)

4.4.1. *rgKF description*

Hereafter, we propose a family of algorithms called regularized Kalman filters or rgKF(OD, E-l^q) that includes an outlier detection (OD) step followed by an l^q ($q > 0$) estimation step (E-l^q). OD stands for the name of the detection method, for example if OD = $NG(\alpha)$, then a non-Gaussian deviation test is applied with a threshold extreme value α.

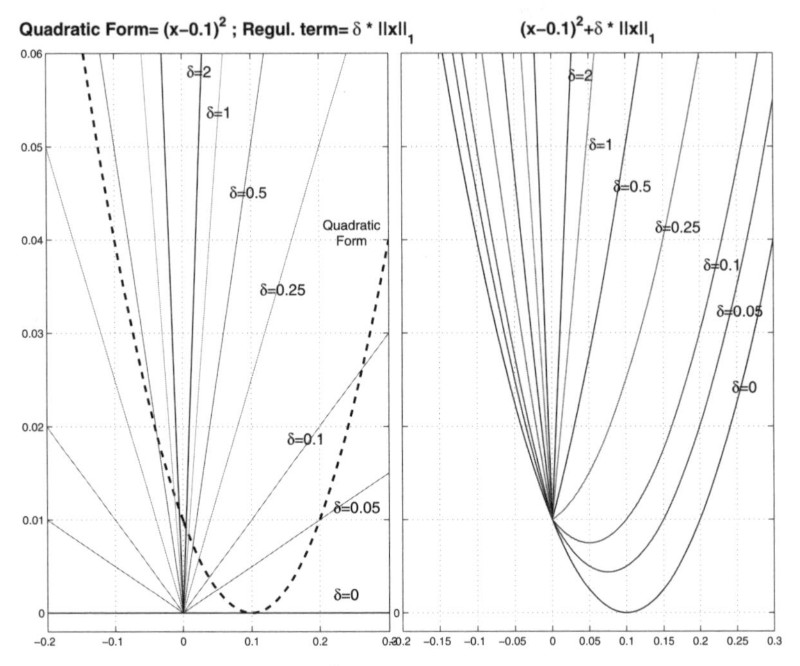

Figure 4.1. *Effect of the l^1-regularization term on the shape of a quadratic form to be minimized*

In the following, we consider only $q = 1, 2$. RKF, which consists of a single and joint detection/estimation step, falls into the l^1-regularization approach and is equivalent to rgKF(l^1, E-l^1). Table 4.1 summarizes these algorithms.

The rgKF is a simple and efficient algorithm. Based on the fact that v_t is an AO in the observations, it is obvious that the innovation $\hat{\epsilon}_{t|t-1}$ in [4.5] should contain v_t if v_t occurs at time t. Moreover, the observation error ϵ_t is assumed to be a Gaussian random variable with a fixed variance σ_ϵ^2. If v_t occurs at time t, then $\hat{\epsilon}_{t|t-1}$ should not lie in the range space of the Gaussian density.

	Outlier detection (OD) method	Estimation method (E-l^q)	Value for δ	$\hat{v}_t$
KF	no	no	no	0
RKF	no	l^1	fixed	$\min_{v_t} C(v_t,\delta)$
rgKF(NG, E-l^1)	$NG(\alpha)$	l^1	$\delta \in [0, 2\,q_t\,\lvert\hat{\epsilon}_{t\lvert t-1}\rvert]$	$\min_{v_t} C(v_t,\delta)$
rgKF(NG, E-l^2)	$NG(\alpha)$	l^2	$\delta \in \left]0, \dfrac{\alpha\,\sigma_\epsilon\,q_t}{\lvert\hat{\epsilon}_{t\lvert t-1}\rvert - \alpha\,\sigma_\epsilon}\right]$	$\dfrac{\hat{\epsilon}_{t\lvert t-1}\,q_t}{\delta_{t,l^2}+q_t}$

Table 4.1. *Family of outlier detection and estimation algorithms*

Our rgKF algorithm [JAY 11b] with exogenous outlier *detection and estimation* can be summarized in the following steps:

$$\text{Prediction}\quad \hat{\boldsymbol{\theta}}_{t|t-1} = \hat{\boldsymbol{\theta}}_{t-1|t-1}, \qquad [4.12]$$

$$\mathbf{S}_{t|t-1} = \mathbf{S}_{t-1|t-1} + \sigma_w^2\,\mathbf{I}, \qquad [4.13]$$

$$\text{Kalman gain}\quad \mathbf{z}_t = \mathbf{S}_{t|t-1}\,\mathbf{g}_t\,(\mathbf{g}_t'\,\mathbf{S}_{t|t-1}\,\mathbf{g}_t + \sigma_\epsilon^2)^{-1} \qquad [4.14]$$

$$\text{Innovation}\quad \hat{\epsilon}_{t|t-1} = r_t - \mathbf{g}_t'\,\hat{\boldsymbol{\theta}}_{t|t-1}, \qquad [4.15]$$

$$\textit{Mean prediction}\quad t>2{:}\ \hat{\mu}_{t|t-1} = \frac{t-1}{t}\hat{\mu}_{t-1|t-1} + \frac{1}{t}\hat{\epsilon}_{t|t-1},$$

$$\text{with } \hat{\mu}_{2|2} = \hat{\epsilon}_{2|1}, \qquad [4.16]$$

$$\textit{Variance prediction}\quad t>2{:}\ \hat{\sigma}^2_{t|t-1} = \frac{t-2}{t-1}\hat{\sigma}^2_{t-1|t-1}$$

$$+\,t\,(\hat{\mu}_{t|t-1} - \hat{\mu}_{t-1|t-1})^2, \text{ with } \hat{\sigma}^2_{2|2} = 0, \qquad [4.17]$$

Outlier detection non-Gaussian test: [4.18]

$$H_0(\alpha){:}\quad \lvert\hat{\epsilon}_{t|t-1} - \hat{\mu}_{t|t-1}\rvert < \alpha\,\hat{\sigma}_{t|t-1} \text{ (no outlier)}, \qquad [4.19]$$

$$H_1(\alpha){:}\quad \lvert\hat{\epsilon}_{t|t-1} - \hat{\mu}_{t|t-1}\rvert \geq \alpha\,\hat{\sigma}_{t|t-1}, \qquad [4.20]$$

Outlier estimation

$$H_0(\alpha) \text{ verified: } \hat{v}_t = 0, \qquad [4.21]$$

$H_1(\alpha)$ verified: – l^1 case: choose $\delta_t > 0$ next to 0 and $\leq 2\, q_t \, |\hat{\epsilon}_{t|t-1}|$ and search for $\hat{v}_t = \hat{v}_{t,l^1}$:

$$\min_{v_t} (\hat{\epsilon}_{t|t-1} - v_t)' \, q_t \, (\hat{\epsilon}_{t|t-1} - v_t) + \delta_t \, \|v_t\|_1, \quad [4.22]$$

– l^2 case: $C(v_t, \delta) = (\hat{\epsilon}_{t|t-1} - v_t)'$

$$q_t \, (\hat{\epsilon}_{t|t-1} - v_t) + \delta_{t,l^2} \, \|v_t\|_2^2 \quad [4.23]$$

$$\hat{v}_t = \hat{v}_{t,l^2} = \frac{\hat{\epsilon}_{t|t-1} \, q_t}{\delta_{t,l^2} + q_t},$$

$$\text{where } \delta_{t,l^2} \in \left] 0, \frac{\alpha \, \sigma_\epsilon \, q_t}{|\hat{\epsilon}_{t|t-1}| - \alpha \, \sigma_\epsilon} \right], \quad [4.24]$$

Correction $$\mathbf{S}_{t|t} = (\mathbf{I} - \mathbf{z}_t \, \mathbf{g}_t') \, \mathbf{S}_{t|t-1}, \quad [4.25]$$

$$\hat{\boldsymbol{\theta}}_t = \hat{\boldsymbol{\theta}}_{t|t-1} + \mathbf{z}_t \, (\hat{\epsilon}_{t|t-1} - \hat{v}_t), \quad [4.26]$$

Mean correction $$\hat{\mu}_{t|t} = \hat{\mu}_{t|t-1} - \frac{1}{t} \hat{v}_t, \quad [4.27]$$

Variance correction $$\hat{\sigma}^2_{t|t} = \frac{t-2}{t-1} \hat{\sigma}^2_{t-1|t-1} + t \, (\hat{\mu}_{t|t} - \hat{\mu}_{t-1|t-1})^2. \quad [4.28]$$

Equations [4.12]–[4.15], [4.22] and [4.25]–[4.26] are similar to the RKF recursive equations as RKF is a particular case of rgKF.

The OD step [4.18] presented above is a classical non-Gaussian test. It allows for incorporating the recursive prediction and correction equations for the first two moments of the innovation process (mean and variance) and has thus the advantage of keeping the recursive property of the original KF.

Parameter α of the hypothesis tests [4.19] and [4.20] is often around 3 but may be chosen as an α-quantile of the Gaussian distribution. For example, if we aim at detecting an outlier within a 1% confidence interval, then α is such that $\mathbb{P}[|\hat{\epsilon}_{t|t-1} - \hat{\mu}_{t|t-1}|/\hat{\sigma}_{t|t-1} \geq \alpha] = 0.005$, which is $\alpha = 2.5758$ (such values can be found in the statistical tables like in [FOR 10]). The often used value of $\alpha = 3$ corresponds to a double confidence interval of 0.13%.

In the l^1-regularization case (equation [4.22]), and if an outlier is detected (if the OD-test $H_1(\alpha)$ is verified), then the value for δ ranges from 0 to $2\,q_t\,|\hat{\epsilon}_{t|t-1}|$. In practice, δ should be close to 0 and will never tend to its upper bound, which is very generous. As shown in Figure 4.1, the smaller is δ and the less regularized is the quadratic form, which allows for estimating potential extreme values. On the contrary, a high value of δ reduces the outlier value to zero, extending therefore the support of the underlying distribution. The OD step bypasses the tricky problem of choosing a convenient value for the regularization parameter.

In the l^2-regularization case, the estimated value for the outlier still depends on the chosen value for δ. The l^2-regularization is often preferred in the literature when closed-form expressions of [4.23] need to be derived (like in [4.24]) and to insure good mathematical properties such as a differentiable and continuous function. In this section, we are mainly interested in the l^1-regularization because of the sparsity of the additive term v_t that is supposed to represent spiky returns.

4.4.2. *rgKF performance*

To demonstrate why rgKF should be applied in any case, this section discusses the performance of the rgKF-l^1 and l^2 on simulated data intentionally corrupted by sparse outliers. The simulated data are similar to the data presented in [3.51], that is a two-factor model (including the S&P500 Equity Index and the T-Bond 10y) and piecewise constant parameters w_t and $(1 - w_t)$. The difference is that we use the end-of-month returns of the two factors in this example (instead of daily data) to match the sampling frequency of the Fung and Hsieh factors and the hedge fund returns used in the next section. Spikes are then intentionally added to the observations and are generated according to a Bernoulli–Gaussian process occurring at five random instant times in the time series. When occurring, the spikes are sampled from a Gaussian density of variance 0.25, which is approximately 10 times higher than the average variance of the two factors (the average daily numbers of Table 3.1 times $\sqrt{12}$ – to get monthly equivalent numbers – compared to $\sqrt{0.25} = 0.5$ give a ratio of around 10).

Figure 4.2 illustrates and compares the performance of the KF and the rgKF, when an NG(3)-test is applied followed by, respectively, an l^2-and an l^1-regularization. We set $\delta_{l^2} = 0.5$ and $\delta_{l^1} = 0.01$. Left-hand side graphs are for the non-spiky data and the right-hand side graphs are for the spiky data. The results are obvious: KF delivers the most reliable estimates when no outlier is added and the rgKF-l^1 results are very close to the KF results. When outliers appear, KF fails to track the parameters whereas rgKF shows good performance. As expected, rgKF-l^2 will depend on the fixed value of δ: when δ is small, then a too large set of data is considered as outlier; when δ increases, outliers are not regularized and perturb the estimation of the parameters. In this case, $\delta_{l^2} = 0.5$ seemed to be a good intermediate trade-off and we decided to show only one example for the sake of clarity. On its side, rgKF-l^1 does very well and is close to the KF results.

In conclusion, rgKF-l^1 should be a substitute for any other method in estimating the dynamic parameters of factor models. Generally, we do not have so much *a priori* information about the presence of outliers. KF already exhibits better tracking abilities than sliding window ordinary least squares (SW-OLS) but it will always lead to spurious results if at least one outlier or extreme event occurs. If the outliers are not filtered from the recursive equations, then what should be considered as specific events might be considered as systematic events, which is obviously misleading.

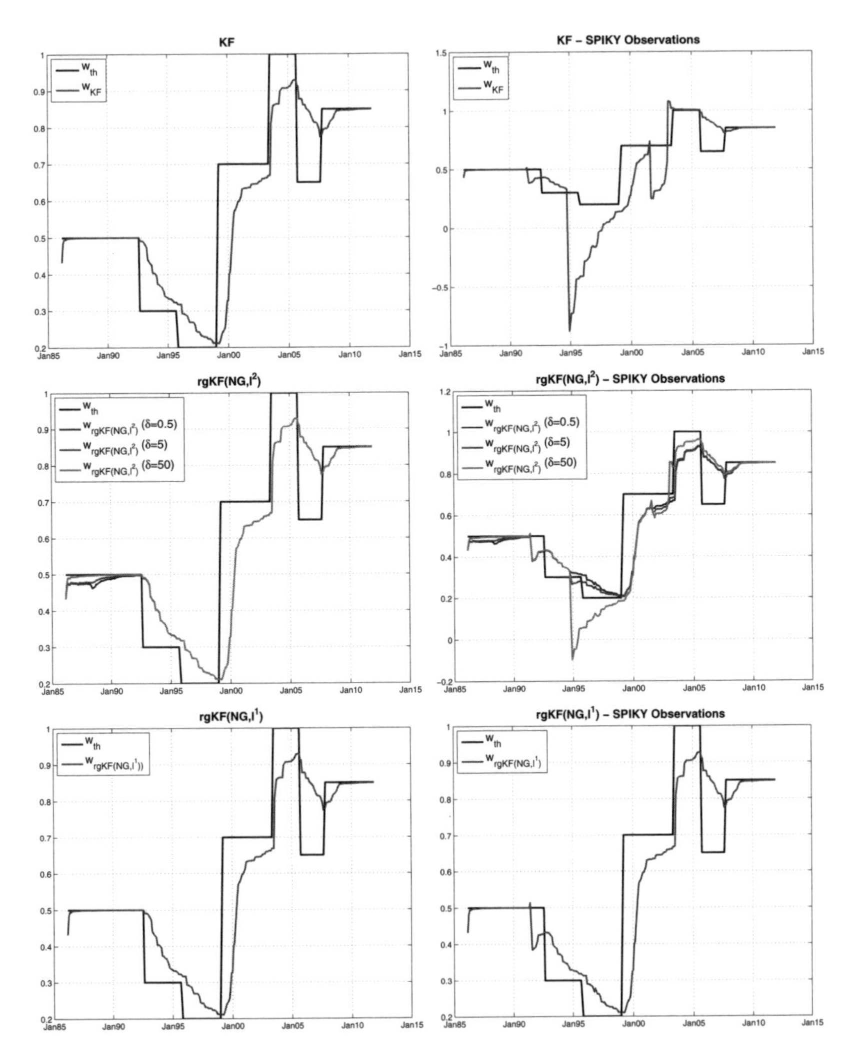

Figure 4.2. *Performance of KF, rgKF-l^2 and rgKF-l^1 on synthetic monthly returns using a two-factor model (S&P500 and US TBond 10y) and piecewise constant parameters (see model [3.51] applied on monthly returns). On the right-hand side graphs, spikes are randomly added in the observations. rgKF-l^1 is therefore the most adapted algorithm*

4.5. Application to detect irregularities in hedge fund returns

The goal of our example is to explain the importance of removing outliers using rgKF algorithms. A multi-factor decomposition of the returns helps us to understand which part of the returns comes from the market and which part is specific (as shown in Table 4.2). If outliers appear in the observations and if the model does not take them into account, then market exposure estimates can be totally spurious. rgKF algorithms avoid the contamination of these estimates by simply filtering outliers. In practice, portfolio managers may use multi-factor approaches to hedge one specific or all market risks. Bad estimates can lead to some inappropriate decisions being made concerning the optimal hedging portfolio.

	Systematic term	Idiosyncratic term
Asset returns	Market returns	Asset-specific returns
r_t	$\mathbf{f}_t' \boldsymbol{\beta}_t$	$\alpha_t + v_t + \epsilon_t$
Risks	Market risks	Idiosyncratic risks
$Var(r_t)$	$\boldsymbol{\beta}_t' \boldsymbol{\Sigma}_f \boldsymbol{\beta}_t$	$\sigma_v^2 + \sigma_\epsilon^2$

Table 4.2. *Breakdown of asset returns and risks into systematic and idiosyncratic components. $\boldsymbol{\Sigma}_f$ is the covariance matrix of the factors*

This section deals with applying rgKF(NG, l^1) on real hedge fund data to detect the potential occurrence of outliers. We conduct the analysis on returns of *Global macro* hedge funds. Managers of such funds take directional positions in currencies, debt, equities or commodities and may also elect to take relative positions combining long positions paired off against short positions. Hedge fund data come from the hedge fund research (HFR) database [RES]. The returns are end-of-month returns, available from January 1998 to June 2010 (i.e. for 150 months).

We choose the eight risk factors of Fung and Hsieh (F&H) [HSI], selected and constructed by the authors [FUN 01] to explain trend-follower hedge funds. These factors are now widely used in the literature of hedge fund analysis. They are quoted on a monthly basis and stored in f_t at each end-of-month t. The reader may refer to section 2.2.3 for a more detailed description of these factors.

Figure 4.3 illustrates the capability of the rgKG(NG,l^1) to detect, estimate and filter some points in the observed returns time series, which might be considered as a hedge fund's specific events, outliers, irregularities or illiquidity issues. As can be seen on the top graph in Figure 4.3, when an outlier occurs, then the resulting KF estimations of β suddenly jump to four or five times their previous values whereas no special event occurs in the market.

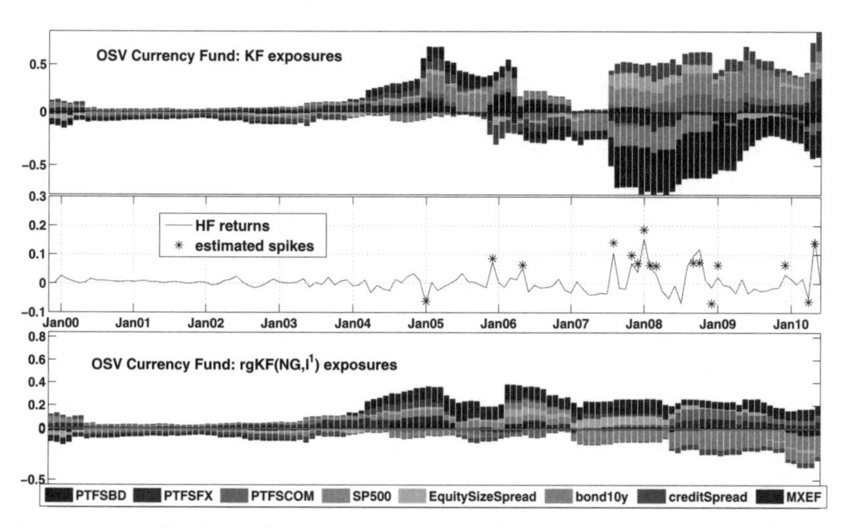

Figure 4.3. *Outliers detection and estimation for a single trend-follower hedge funds of the database HFR (OSV Currency Fund). Impact of outliers on the fund's exposures can be observed on the top graph: KF does not filter the rgKF(NG,l^1) estimated spikes and gives incorrect estimations of the market risks (see color plate section)*

4.6. Conclusion

This chapter has presented a new family of algorithms named rgKFs that have been derived to *detect and estimate* exogenous outliers that might occur in the observation equation of a standard KF. Inspired from the RKF of Mattingley and Boyd [MAT 10], which makes use of a l^1-regularization step, we introduce a simple but efficient detection step in the recursive equations of the RKF. This solution is one means by which to solve the problem of adapting the value of the l^1-regularization parameter: when an outlier is detected in the innovation term of the KF, the value of the regularization parameter is set to a value that will let the l^1-based optimization problem estimate the amplitude of the spike. We have also tested an l^2-based regularization term for which the solution is known in a closed form.

The results obtained on synthetic data show that the best performance is achieved by the rgKF(NG,l^1) that handles the choice of the regularization parameter. In the context of analyzing hedge funds returns, the rgKF(NG,l^1) could help risk managers to dynamically and accurately estimate the exposures of hedge funds for which they do not have any other information except the observed returns. Interpreting outliers as illiquidity issues, our algorithms would therefore be very useful to classify funds according to their "illiquidity quality" and to make accurate decisions from a regulation point of view.

4.7. Chapter highlights

Extended space-state model

$$\boldsymbol{\theta}_t = \boldsymbol{\theta}_{t-1} + \mathbf{w}_t,$$

$$r_t = \mathbf{g}_t' \boldsymbol{\theta}_t + v_t + \epsilon_t.$$

As for the classical KF state-space model, $\mathrm{w} \sim \mathcal{N}(\mathbf{0}, \sigma_w^2 \mathbf{I})$ and $\epsilon \sim \mathcal{N}(0, \sigma_{\epsilon^2})$. Additional v_t is a stochastic impulse process, for example a Bernoulli–Gaussian process.

Robust Kalman filter (RKF)

The RKF algorithm consists of the following steps:

$$
\begin{aligned}
\text{Prediction} \quad & \hat{\boldsymbol{\theta}}_{t|t-1} = \hat{\boldsymbol{\theta}}_{t-1|t-1}, \\
& \mathbf{S}_{t|t-1} = \mathbf{S}_{t-1|t-1} + \sigma_w^2 \mathbf{I}, \\
\text{Innovation} \quad & \hat{\epsilon}_{t|t-1} = r_t - \mathbf{g}_t' \hat{\boldsymbol{\theta}}_{t|t-1}, \\
\text{Kalman gain} \quad & \mathbf{z}_t = \mathbf{S}_{t|t-1} \mathbf{g}_t (\mathbf{g}_t' \mathbf{S}_{t|t-1} \mathbf{g}_t + \sigma_\epsilon^2)^{-1}, \\
\text{Correction} \quad & \mathbf{S}_{t|t} = (\mathbf{I} - \mathbf{z}_t \mathbf{g}_t') \mathbf{S}_{t|t-1}, \\
& \hat{\boldsymbol{\theta}}_t = \hat{\boldsymbol{\theta}}_{t|t-1} + \mathbf{z}_t (\hat{\epsilon}_{t|t-1} - \hat{v}_t), \\
\text{Regularization} \quad & \hat{v}_t \text{ is the solution of } \min_{v_t} (\hat{\epsilon}_{t|t-1} - v_t)' \\
& q_t (\hat{\epsilon}_{t|t-1} - v_t) + \delta \|v_t\|_1 = \min_{v_t} C(v_t, \delta), \text{with} \\
& q_t = (1 - \mathbf{g}_t' \mathbf{z}_t)' \sigma_\epsilon^{-2} (1 - \mathbf{g}_t' \mathbf{z}_t) + \mathbf{z}_t' \mathbf{S}_{t|t-1}^{-1} \mathbf{z}_t, q_t > 0 \\
& \text{and } \delta \geq 0 \text{ is the regularization parameter.}
\end{aligned}
$$

Regularized Kalman filter (rgKF)

Our *rgKF* algorithm [JAY 11b] with exogenous outlier *detection and estimation* can be summarized in the following steps:

$$
\begin{aligned}
\text{Prediction} \quad & \hat{\boldsymbol{\theta}}_{t|t-1} = \hat{\boldsymbol{\theta}}_{t-1|t-1}, \\
& \mathbf{S}_{t|t-1} = \mathbf{S}_{t-1|t-1} + \sigma_w^2 \mathbf{I}, \\
\text{Kalman gain} \quad & \mathbf{z}_t = \mathbf{S}_{t|t-1} \mathbf{g}_t (\mathbf{g}_t' \mathbf{S}_{t|t-1} \mathbf{g}_t + \sigma_\epsilon^2)^{-1}, \\
\text{Innovation} \quad & \hat{\epsilon}_{t|t-1} = r_t - \mathbf{g}_t' \hat{\boldsymbol{\theta}}_{t|t-1},
\end{aligned}
$$

Mean prediction $t > 2: \hat{\mu}_{t|t-1} = \dfrac{t-1}{t}\hat{\mu}_{t-1|t-1} + \dfrac{1}{t}\hat{\epsilon}_{t|t-1},$

with $\hat{\mu}_{2|2} = \hat{\epsilon}_{2|1},$

Variance prediction $t > 2: \hat{\sigma}^2_{t|t-1} = \dfrac{t-2}{t-1}\hat{\sigma}^2_{t-1|t-1}$

$+t\left(\hat{\mu}_{t|t-1} - \hat{\mu}_{t-1|t-1}\right)^2,$ with $\hat{\sigma}^2_{2|2} = 0,$

Outlier detection Non-Gaussian test:

$H_0(\alpha)$: $\left|\hat{\epsilon}_{t|t-1} - \hat{\mu}_{t|t-1}\right| < \alpha\,\hat{\sigma}_{t|t-1}$ (no outlier),

$H_1(\alpha)$: $\left|\hat{\epsilon}_{t|t-1} - \hat{\mu}_{t|t-1}\right| \geq \alpha\,\hat{\sigma}_{t|t-1},$

Outlier estimation

$H_0(\alpha)$ verified: $\hat{v}_t = 0,$

$H_1(\alpha)$ verified: – l^1 case: choose $\delta_t > 0$ next to 0 and

$\leq 2\,q_t\left|\hat{\epsilon}_{t|t-1}\right|$ and search for

$\hat{v}_t = \hat{v}_{t,l^1}: \min_{v_t}(\hat{\epsilon}_{t|t-1} - v_t)'$

$q_t\left(\hat{\epsilon}_{t|t-1} - v_t\right) + \delta_t\left\|v_t\right\|_1,$

– l^2 case: $C(v_t, \delta) = \left(\hat{\epsilon}_{t|t-1} - v_t\right)' q_t\left(\hat{\epsilon}_{t|t-1} - v_t\right)$

$+\delta_{t,l^2}\left\|v_t\right\|_2^2$

$\hat{v}_t = \hat{v}_{t,l^2} = \dfrac{\hat{\epsilon}_{t|t-1}\,q_t}{\delta_{t,l^2} + q_t},$

where $\delta_{t,l^2} \in \left]0, \dfrac{\alpha\,\sigma_\epsilon\,q_t}{\left|\hat{\epsilon}_{t|t-1}\right| - \alpha\,\sigma_\epsilon}\right],$

Correction $\mathbf{S}_{t|t} = \left(\mathbf{I} - \mathbf{z}_t\,\mathbf{g}_t'\right)\mathbf{S}_{t|t-1},$

$\hat{\boldsymbol{\theta}}_t = \hat{\boldsymbol{\theta}}_{t|t-1} + \mathbf{z}_t\left(\hat{\epsilon}_{t|t-1} - \hat{v}_t\right),$

Mean correction $\hat{\mu}_{t|t} = \hat{\mu}_{t|t-1} - \dfrac{1}{t}\hat{v}_t,$

Variance correction $\hat{\sigma}^2_{t|t} = \dfrac{t-2}{t-1}\hat{\sigma}^2_{t-1|t-1} + t\left(\hat{\mu}_{t|t} - \hat{\mu}_{t-1|t-1}\right)^2.$

Appendix

Some Probability Densities

A1.1. Gaussian distribution

A1.1.1. *Gaussian vector (with a multivariate Gaussian density)*

If an N-dimensional random vector x is Gaussian distributed, then its probability density is fully characterized by its two first centered moments, its mean $\boldsymbol{\mu} = \mathbb{E}[\mathrm{x}]$ and its full-rank (invertible) covariance matrix $\boldsymbol{\Sigma} = \mathbb{E}[(\mathrm{x} - \boldsymbol{\mu})(\mathrm{x} - \boldsymbol{\mu})^T]$ whose expression is as follows:

$$p(\mathrm{x}) = \frac{1}{(2\pi)^{N/2} |\boldsymbol{\Sigma}|^{1/2}} \exp\left(-\frac{(\mathrm{x} - \boldsymbol{\mu})^T \boldsymbol{\Sigma}^{-1} (\mathrm{x} - \boldsymbol{\mu})}{2} \right), \quad [A1.1]$$

where $|\boldsymbol{\Sigma}|$ denotes the determinant of the covariance matrix $\boldsymbol{\Sigma}$. It is usually denoted by $\mathrm{x} \sim \mathcal{N}(\boldsymbol{\mu}, \boldsymbol{\Sigma})$.

A1.1.2. *White Gaussian vector with iso-variance components*

The vector x is said to be "white" if its N components $x_k, 1 \leq k \leq N$ are independent. This implies that,

$\mathbb{E}[x_k x_m] = \mathbb{E}[x_k]\mathbb{E}[x_m]$, $k \neq m$. Independent random variables are consequently uncorrelated, that is $\mathbb{E}[\underline{x}_k \underline{x}_m] = 0$, $k \neq m$. If a random vector is Gaussian and white, with all its components sharing the same variance σ^2, (i.e. $\mathbb{E}[\underline{x}_k^2] = \sigma^2$, $1 \leq k \leq N$), then its covariance matrix is proportional to the $N \times N$ identity matrix, $\mathbf{\Sigma} = \sigma^2 \mathbf{I}_N$ and the form [A1.1] simplifies into:

$$p(\mathrm{x}) = \frac{1}{(2\pi)^{N/2}\sigma^N} \exp\left(-\frac{(\mathrm{x}-\boldsymbol{\mu})^T(\mathrm{x}-\boldsymbol{\mu})}{2\sigma^2}\right). \quad \text{[A1.2]}$$

Equation [A1.2] displays a "parsimonious" Probability Density (in the sense of section 3.8.3) that depends only on $N+1$ parameters, that is the N means, $\mathbb{E}[x_k] = \mu_k$, $1 \leq k \leq N$ and the variance σ^2. Moreover, if x is zero mean, that is $\boldsymbol{\mu} = \mathbb{E}[\mathrm{x}] = 0$, then $p(\mathrm{x})$ depends solely upon the variance σ^2.

A1.1.3. *Gaussian random variable*

If $N = 1$, then [A1.1] becomes the Probability Density of a Gaussian random variable x, that is $x \sim \mathcal{N}(\mu, \sigma^2)$, with:

$$p(x) = \frac{1}{\sigma\sqrt{2\pi}} \exp\left(-\frac{(x-\mu)^2}{2\sigma^2}\right). \quad \text{[A1.3]}$$

A1.2. χ^2 distribution

A1.2.1. χ^2 *probability density*

A random variable x is χ^2-distributed with ν degree of freedom (or $x \sim \chi^2(\nu)$) if, and only if, its probability density function is given by, $\forall x > 0$:

$$p_\nu(x) = \frac{1}{2^{\nu/2}\Gamma(\nu/2)} x^{\nu/2-1} e^{-x/2}, \quad \text{[A1.4]}$$

where $\Gamma(.)$ is the gamma function defined for all complex number z by:

$$\Gamma(z) = \int_0^{\infty} t^{z-1} e^{-t} \, dt. \quad \text{[A1.5]}$$

In particular, we have $\mathbb{E}(x) = \nu$ and $Var(x) = 2\nu$.

A1.2.2. *A χ^2 variable is the sum of squared i.i.d. Gaussian variables*

If we consider N i.i.d. Gaussian random variables x_i, with mean μ_i and variance σ_i, for $i = 1, \cdots, N$, then:

$$y = \sum_{i=1}^{N} \left(\frac{x_i - \mu_i}{\sigma_i} \right)^2 \sim \chi^2(N). \quad \text{[A1.6]}$$

Following the central limit theorem, as soon as N is large ($N > 100$), the distribution of the random variable y may be approximated by a Gaussian distribution with mean N and variance $2N$.

A1.3. Student's *t*-distribution

A1.3.1. *Student's t-probability density*

The Student's t-distribution of a random variable t is characterized by a single parameter $\nu \geq 0$ and is given by:

$$p_\nu(t) = \frac{1}{\sqrt{\pi \nu}} \frac{\Gamma(\frac{\nu+1}{2})}{\Gamma(\nu/2)} \left(1 + \frac{t^2}{\nu} \right)^{-\frac{\nu+1}{2}}. \quad \text{[A1.7]}$$

It is usually denoted as $t \sim t(\nu)$. In particular, we have $\mathbb{E}(t) = 0$ for $\nu > 1$, $\mathbb{E}(t)$ undefined for $\nu = 1$ and $Var(t) = \nu/(\nu - 2)$ if $\nu > 2$, infinite for $\nu \leq 2$. The function $\Gamma(.)$ is given in equation [A1.5].

A1.3.2. *A Student's t variable is the ratio between a Gaussian-distributed variable and a χ^2-distributed variable*

If $x \sim \mathcal{N}(0,1)$ and $u \sim \chi^2(\nu)$, then:

$$t = \frac{x}{\sqrt{u/\nu}} \sim t(\nu). \qquad \text{[A1.8]}$$

A1.3.3. *Confidence interval of the estimate of the mean of a Gaussian random variable with unknown variance*

The confidence interval at α-confidence level for $\hat{\mu}$, an estimate of the mean value μ of x (see [A1.3] with σ^2 unknown and estimated by s^2), is given by:

$$\left[\hat{\mu} - q_{1-\alpha/2}^{t(N-1)} \sqrt{\frac{s^2}{N}}; \hat{\mu} + q_{1-\alpha/2}^{t(N-1)} \sqrt{\frac{s^2}{N}}\right], \qquad \text{[A1.9]}$$

where $q_{1-\alpha/2}^{t(N-1)}$ is the $(1 - \alpha/2)$ quantile of the Student's t variable $t(N-1)$.

A1.4. Fisher (or Fisher–Snedecor or Snedecor's *F*) distribution

A1.4.1. *The Fisher (or Fisher–Snedecor, or Snedecor's F) probability density*

This probability density $F(p,q)$ is given, $\forall x \in \mathbb{R}^+$ and $p, q \in \mathbb{N}^*$, by:

$$p(x) = \frac{\left(\frac{px}{px+q}\right)^{p/2}\left(1 - \frac{px}{px+q}\right)^{q/2}}{x\,\mathcal{B}(p/2, q/2)}, \qquad \text{[A1.10]}$$

where $\mathcal{B}(a,b)$ is the beta function defined by:

$$\mathcal{B}(a,b) = \int_0^1 t^{a-1}\,(1-t)^{b-1}\,dt. \qquad \text{[A1.11]}$$

In particular, we have $\mathbb{E}(F) = \frac{\nu_2}{\nu_2-2}$ for $\nu_2 > 2$, and $\mathbb{E}(F) = \dfrac{2\,\nu_2^2\,(\nu_1+\nu_2-2)}{\nu_1\,(\nu_2-2)^2\,(\nu_2-4)}$ for $\nu_2 > 4$.

A1.4.2. *A Fisher variable is the ratio between two χ^2 variables*

If $u_1 \sim \chi^2(\nu_1)$ and $u_2 \sim \chi^2(\nu_2)$, then:

$$y = \frac{u_1/\nu_1}{u_2/\nu_2} \sim F(\nu_1,\nu_2). \qquad \text{[A1.12]}$$

A1.4.3. *The inverse of a Fisher variable is a Fisher variable*

If $u \sim F(\nu_1,\nu_2)$, then $\dfrac{1}{u} \sim F(\nu_2,\nu_1)$.

A1.4.4. *A squared Fisher variable may be a Student's-t variable*

If $u \sim t(\nu)$, then $u^2 \sim F(1,\nu)$.

A1.4.5. *A squared Fisher variable may be a Gaussian variable*

If $u \sim \mathcal{N}(0,1)$, then $u^2 \sim F(1,\infty)$.

A1.5. Hotelling's *T*-squared distribution

The Hotelling's T-squared probability density with parameters p and q and denoted by $\mathcal{T}^2(p,q)$ is related to the Fisher probability density [A1.10] as follows: if $u \sim \mathcal{T}^2(p,q)$, then $\frac{q-p+1}{pq} u \sim F(p, q-p+1)$.

A1.6. Wishart distribution

The Wishart distribution denoted by $\mathcal{W}(K, \mathbf{A})$ is the distribution of:

$$\sum_{k=1}^{K} \mathbf{z}_k \mathbf{z}_k', \qquad [A1.13]$$

where the $\mathbf{z}_k$s are independent, identically distributed (i.i.d.) according to a Gaussian distribution with zero mean and covariance $\mathbf{A}$.

A1.7. Marčenko–Pastur distribution

Suppose that we use the sample covariance matrix computed on the standardized matrix of returns $\tilde{\mathrm{R}}$ as described in equation [2.4]. The resulting estimated covariance matrix S therefore has random elements with an assumed variance σ^2. Then, in the limit $T, N \to \infty$, keeping the ratio $q = T/N \geq 1$ constant, the density of the eigenvalues of S is given by:

$$\rho_q(\lambda) = \frac{q}{2\pi\sigma^2} \frac{\sqrt{(\lambda_+ - \lambda)^+ (\lambda - \lambda_-)^+}}{\lambda}, \qquad [A1.14]$$

where the maximum and the minimum eigenvalues are given by:

$$\lambda_{\pm} = \sigma^2 \left(1 \pm \sqrt{\frac{1}{q}}\right)^2 \quad \text{[A1.15]}$$

and $(x - a)^+ = max(0, x - a)$. The density $\rho(\lambda)$ is known as the Marčenko–Pastur density [MAR 67, PAS 11].

A1.8. Bernoulli–Gaussian distribution

A Bernouilli–Gaussian process x is such that given a set S of indices drawn from a set $I = \{1, \cdots, N\}$ with a probability $p \ll 1$, each element of x is identically zero if the corresponding index is not in the set S, otherwise the element is Gaussian with mean μ and non-zero variance σ^2. Elements of x are i.i.d. given the support set.

The probability that the cardinality of the support set S equals K is given by $\mathbb{P}(|S| = K) = \binom{N}{K} p^K (1 - p)^{N-K}$. If $\mathbf{x}_S$ denotes the vector consisting of the elements of x whose indices are in the set S, then the vector $\mathbf{x}_S$ follows i.i.d. Gaussian distribution, that is $\mathbf{x}_S \sim \mathcal{N}(\mu \mathbb{1}_{|S|}, \sigma^2 \mathbf{I}_{|S|})$. Thus, S is the support set of the signal vector x with expected cardinality $\mathbb{E}(|S|) = Np \ll N$ and x is sparse with high probability.

Conclusion

As mentioned in the introduction, our objective in this book is to adapt the QUANT(ITATIVE) tools to the new financial world characterized by noisy data, enhanced risk controls and an increasing demand for more transparency. Throughout this book, the classic linear factor model is our benchmark to prove how the statistical signal processing (SSP) techniques can help in solving this task. Keeping a "user friendly" framework that gives an intuitive explanation of most of the stylized facts empirically observed, we show that more complex phenomena, such as nonlinearities or liquidity issues, can be taken into account to obtain robust results.

Of course, this robust treatment has a cost in terms of complexity. Kalman filter estimators must replace least squares estimators; regularization techniques have to be implemented to treat the noisy impact of illiquidity on asset returns. But SSP gives us a proven technical framework, which is easy to adapt in our context. We have shown in the different chapters how factor selection and parameter estimation can be done to get these robust results that we are targeting.

Most of our empirical illustrations come from the asset management industry, and in particular the hedge fund industry. Linear factor models and style analysis could in theory give a good answer, as is the case for mutual funds. However, hedge fund returns, and their particularities, nonlinearity in particular, encourages us to adapt the basic linear setup by considering either nonlinear factors or time-varying linear exposures to risk factors. The first approach has a main drawback. By introducing nonlinearities in the model, we answer to the demand for transparency by using tools that cannot be considered as "usual practices" in the industry. The second approach gives a more "user friendly" solution as it corresponds to an intuitive extension of the current practices.

Bibliography

[AGA 00a] AGARWAL V., NAIK N., "Multi-period performance persistence analysis of hedge funds", *Journal of Financial and Quantitative Analysis*, vol. 35, no. 3, pp. 327–342, September 2000.

[AGA 00b] AGARWAL V., NAIK N., "On taking the alternative route: risks, rewards and performance persistence of hedge funds", *Journal of Alternative Investments*, vol. 2, no. 4, pp. 6–23, 2000.

[AGA 04] AGARWAL V., NAIK N., "Risks and portfolio decisions involving hedge funds", *Review of Financial Studies*, vol. 17, no. 1, pp. 63–98, 2004.

[AGA 09] AGARWAL V., BAKSHI G.S., HUIJ J., "Do higher-moment equity risks explain hedge fund returns?", *1st Annual Conference on Econometrics of Hedge Funds*, Research Paper No. RHS 06-066, R. H. Smith School, 2009.

[AHN 09] AHN S.C., HORENSTEIN A.R., Eigenvalue ratio test for the number of factors, Mimeo, Arizona State University, 2009.

[AKA 74] AKAIKE H., "A new look at the statistical model identification", *IEEE Transaction on Automatic and Control*, vol. 19, no. 6, pp. 716–723, June 1974.

[AMM 10] AMMANN M., HUBER R.O., SCHMID M.M., "Benchmarking hedge funds: the choice of the factor model", August 2011.

[AND 63] ANDERSON T.W., "Asymptotic theory for principal component analysis", *Annals of Mathematics and Statistics*, vol. 34, pp. 122–148, February 1963.

[AND 97] ANDRIEU C., DOUCET A., DUVAUT P., "Bayesian estimation of filtered point processes using Markov chain Monte Carlo methods", *31st Asilomar Conference on Signals, Systems & Computers*, vol. 2, pp. 1097–1100, November 1997.

[BAI 02] BAI J., NG S., "Determining the number of factors in approximate factor models", *Econometrica*, vol. 70, no. 1, pp. 191–221, 2002.

[BAI 10] BAI Z., SILVERSTEIN J.W., *Spectral Analysis of Large Dimensional Random Matrices*, 2nd. ed., Springer, 2010.

[BAR 93] BAR-SHALOM Y., LI X.R., *Estimation and Tracking: Principles, Techniques, and Software*, Artech House, 1993.

[BIL 12] BILLIO M., GETMANSKY M., PELIZZON L., "Dynamic risk exposure in hedge funds", *Computational Statistics and Data Analysis*, vol. 56, pp. 3517–3532, 2012.

[BIL 99] BILODEAU M., BRENNER D., *Theory of Multivariate Statistics*, Springer, 1999.

[BLA 72] BLACK F., "Capital market equilibrium with restricted borrowing", *Journal of Business*, vol. 45, no. 3, pp. 444–455, 1972.

[BOL 09] BOLLEN N.P.B., WHALEY R.E., "Hedge fund risk dynamics: implications for performance appraisal", *Journal of Finance*, vol. 64, no. 2, pp. 987–1037, 2009.

[BRO 08] BRODIE J., DAUBECHIES I., MOL C.D., *et al.*, Sparse and stable markowitz portfolios, European Central Bank working paper no. 936, September 2008.

[BUR 09] BURASCHI A., KOSOWKI R., TROJANI F., "When there is no place to hide: correlation risk and the cross-section of hedge fund returns", *1st Annual Conference on Econometrics of Hedge Funds*, Imperial College working paper, 2009.

[BUR 03] BURMEISTER E., ROLL R., ROSS S.A., Using macroeconomic factors to control portfolio risk, Working paper, BIRR Portfolio Analysis, Inc., 2003.

[CAM 97] CAMPBELL J., LO A.W., MACKINLAY A.C., *The Econometrics of Financial Markets*, Princeton University Press, 1997.

[CAR 97] CARHART M.M., "On persistence in mutual fund performance", *The Journal of Finance*, vol. 52, no. 1, pp. 57–82, March 1997.

[CAT 66] CATTELL R.B., "The scree test for the number of factors", *Multivariate Behavioral Research*, vol. 1, pp. 245–276, 1966.

[CAU 02] CAUCHIE S., HOESLI M., ISAKOV D., The determinants of stock returns: an analysis of industrial sector indices, Report, HEC-University of Geneva, 2002.

[CHA 83] CHAMBERLAIN G., ROTHSCHILD M., "Arbitrage, factor structure and mean-variance analysis on large asset markets", *Econometrica*, vol. 51, no. 5, pp. 1281–1304, 1983.

[CHA 98] CHAN K.C., KARCESKI J., LAKONISHOK J., "The risk and return from factors", *Journal of Financial and Quantitative Analysis*, vol. 33, no. 2, pp. 159–188, 1998.

[CHE 86] CHEN N.F., ROLL R., ROSS S.A., "Economic forces and the stock market", *Journal of Business*, vol. 59, no. 3, pp. 383–403, 1986.

[CHE 93] CHEN S.J., JORDAN B.D., "Some empirical tests in the arbitrage pricing theory: macrovariables vs. derived factors", *Journal of Banking and Finance*, vol. 17, no. 1, pp. 65–89, 1993.

[CON 95] CONNOR G., "The three types of factor models: a comparison of their explanatory power", *Financial Analysts Journal*, pp. 42–46, May/June 1995.

[CON 02] CONTE E., MAIO A.D., RICCI G., "Recursive estimation of the covariance matrix of a compound-Gaussian process and its application to adaptive CFAR detection", *IEEE Transactions of Signal Processing*, vol. 50, no. 8, pp. 1908–1915, August 2002.

[CRA 46] CRAMER H., *Mathematical Methods of Statistics*, Princeton University Press, 1946.

[CRA 79] CRAVEN P., WAHBA G., "Smoothing noisy data with spline functions", *Numerische Mathematik*, vol. 31, pp. 377–403, 1979.

[CRI 10] CRITON, G., SCAILLET O., "Time-varying analysis in risk and hedge fund performance: how forecast ability increases estimated alpha", 2010.

[CRO 05] CROWLEY P.M., LEE J., "Decomposing the co-movement of the business cycle: a time-frequency analysis of growth cycles in the euro zone", *Macroeconomics 0503015*, EconWPA, pp. 1075–1080, 2005.

[DAR 10] DAROLLES S., GOURIÉROUX C., "Conditional fitted Sharpe performance with application to hedge fund rating", *Journal of Banking and Finance*, vol. 34, no. 3, pp. 578–593, 2010.

[DAR 11] DAROLLES S., MERO G., "Hedge fund returns and factor models: a cross section approach", *Bankers, Markets and Investors*, vol. May–June, no. 112, pp. 34–53, 2011.

[DAR 12] DAROLLES S., VAISSIÉ M., "The alpha and omega of fund of hedge fund added value", *Journal of Banking and Finance*, vol. 36, no. 4, pp. 1067–1078, April 2012.

[DEM 09] DEMIGUEL V., GARLAPPI L., NOGALES F.J., *et al.*, "A generalized approach to portfolio optimization: improving performance by constraining portfolio norms", *Management Science*, vol. 55, no. 5, pp. 798–812, May 2009.

[DJU 03] DJURIĆ P.M., KOTECHA J.H., ZHANG J., *et al.*, "Particle filtering", *IEEE Signal Processing Magazine*, vol. 20, pp. 19–38, September 2003.

[DRA 09] DRAKAKIS K., "Application of signal processing to the analysis of financial data", *IEEE Signal Processing Magazine*, vol. 26, no. 5, pp. 156–160, September 2009.

[DUV 94] DUVAUT P., *Traitement du Signal, Concepts et Applications*, Hermes Science, 1994.

[ELD 05] ELDERMAN A., RAO N.R., "Random matrix theory", *Acta Numerica*, vol. 14, pp. 233–297, 2005.

[FAM 73] FAMA E.F., MACBETH J., "Risk, return and equilibrium: empirical tests", *Journal of Political Economy*, vol. 81, pp. 607–636, 1973.

[FAM 92] FAMA E.F., FRENCH K.R., "The cross-section of expected stock returns", *Journal of Finance*, vol. 47, no. 2, June 1992.

[FAM 93] FAMA E.F., FRENCH K.R., "Common risk factors in the returns on stocks and bonds", *Journal of Financial Economics*, vol. 33, no. 1, pp. 3–56, 1993.

[FOR 10] FORBES C., EVANS M., HASTINGS N., *Statistical Distributions*, 4th ed., John Wiley & Sons, 2010.

[FOX 72] FOX A.J., "Outliers in time series", *Journal of the Royal Statistical Society, Series B*, vol. 34, pp. 350–363, 1972.

[FRA 04] FRAHM G., Generalized elliptical distributions: theory and applications, PhD Thesis, University of Koln, July 2004.

[FRE 03] FRENCH C.W., "The Treynor capital asset pricing model", *Journal of Investment Management*, vol. 1, no. 2, pp. 60–72, 2003.

[FUK 90] FUKUNAGA K., *Introduction to Statistical Pattern Recognition*, 2nd ed., Academic Press, 1990.

[FUN 01] FUNG W., HSIEH D., "The risk in hedge fund strategies, theory and evidence from trend followers", *Review of Financial Studies*, vol. 14, no. 2, pp. 313–341, 2001.

[GEN 05] GENÇAY R., SELÇUK F., WHITCHER B., "The risk in hedge fund strategies, theory and evidence from trend followers", *Journal of International Money and Finance*, vol. 24, pp. 55–70, 2005.

[GIN 02] GINI F., GRECO M.V., "Covariance matrix estimation for CFAR detection in correlated heavy tailed clutter", *Signal Processing*, vol. 82, no. 12, pp. 1847–1859, December 2002.

[HAM 94] HAMILTON J.D., *Time Series Analysis*, Princeton University Press, 1994.

[HAR 07] HARDING M.C., Essays in econometrics and random matrix theory, PhD Thesis, Massachusetts Institute of Technology, May 2007.

[HAR 09] HARVEY A., KOOPMAN S.J., "Unobserved components models in economics and finance", *IEEE Control Systems Magazine*, vol. 29, no. 6, pp. 71–81, December 2009.

[HOT 31] HOTELLING H., "The generalization of Student's ratio", *Annals of Mathematical Statistics*, vol. 2, no. 3, pp. 360–378, 1931.

[HSI] HSIEH D. A., "Data library webpage: hedge fund risk factors". Available at http://faculty.fuqua.duke.edu/%7Edah7/HFRFData.htm.

[HUB 64] HUBER P.J., "Robust estimation of a location parameter", *The Annals of Mathematical Statistics*, vol. 35, no. 1, pp. 73–101, 1964.

[HUB 67] HUBER P.J., "The behavior of maximum likelihood estimates under nonstandard conditions", *5th Berkeley Symposium on Mathematical Statistics and Probability*, vol. 1, pp. 221–233, 1967.

[HUB 77] HUBER P.J., *Robust Covariances in Statistical Decision Theory and Related Topics*, Academic Press Inc., New York, 1977.

[JAG 03] JAGANNATHAN R., MA T., "Risk reduction in large portfolios: why imposing the wrong constraints helps", *Journal of Finance*, vol. 58, no. 4, pp. 1651–1684, 2003.

[JAY 03] JAY E., Détection en environnement non-Gaussien, application à la détection radar, PhD Thesis, University of Cergy-Pontoise/ONERA, June 2003.

[JAY 11a] JAY E., DUVAUT P., DAROLLES S., *et al.*, "Multi-factor models: examining the potential of signal processing techniques", *IEEE Signal Processing Magazine*, vol. 28, no. 5, pp. 37–48, September 2011.

[JAY 11b] JAY E., DUVAUT P., DAROLLES S., *et al.*, "Lq-regularization of the Kalman filter for exogenous outlier removal: application to hedge funds analysis", *4th International Workshop on Computational Advances in Multi-Sensor Adaptive Processing (CAMSAP)*, IEEE Signal Processing Society, December 2011.

[KAL 89] KALABA R., TESFATSION L., "Time-varying linear regression via flexible least squares", *Computers and Maths with Applications*, vol. 17, no. 8–9, pp. 1215–1245, 1989.

[KAY 93] KAY S.M., *Fundamentals of Statistical Signal Processing, Volume I: Estimation Theory*, Prentice Hall, 1993.

[KAY 98] KAY S.M., *Fundamentals of Statistical Signal Processing, Volume 2: Detection Theory*, Prentice Hall, 1998.

[KEN 01] KENT D.D., HIRSHLEIFER D., SUBRAHMANYAM A., "Overconfidence, arbitrage, and equilibrium asset pricing", *Journal of Finance*, vol. LVI, no. 3, pp. 921–965, June 2001.

[LIN 65] LINTNER J., "The valuation of risk assets and selection of risky investments in stock portfolios and capital budgets", *Review of Economics and Statistics*, vol. 47, no. 1, pp. 13–37, 1965.

[MAL 73] MALLOWS C.L., "Some comments on Cp", *Technometrics*, vol. 15, pp. 661–675, 1973.

[MAR 67] MARČENKO V.A., PASTUR L.A., "Distribution of eigenvalues for some sets of random matrices", *Mathematics of the USSR Sbornik*, vol. 1, no. 4, pp. 457–483, 1967.

[MAR 52] MARKOWITZ H.M., "Portfolio selection", *Journal of Finance*, vol. 7, no. 1, pp. 77–91, 1952.

[MAR 04] MARKOV M., MOTTL V., MUCHNIK I., Principles of nonstationary regression estimation: a new approach to dynamic multi-factor models in finance, Technical Report number 2004-47, DIMACS, Rutgers University, Piscataway, NJ, October 2004.

[MAR 76] MARONNA R.A., "Robust M-estimators of multivariate location and scatter", *Annals of Statistics*, vol. 4, no. 1, pp. 51–67, 1976.

[MAR 06] MARONNA R.A., MARTIN R.D., YOHAI V.J., *Robust Statistics, Theory and Methods*, John Wiley & Sons, 2006.

[MAT 10] MATTINGLEY J., BOYD S., "Real-time convex optimization in signal processing", *IEEE Signal Processing Magazine*, vol. 27, no. 3, pp. 50–61, May 2010.

[MOS 66] MOSSIN J., "Equilibrium in a capital market", *Econometrica*, vol. 34, no. 4, pp. 768–783, 1966.

[NOR 04] NORMAN S., *Control System Engineering*, John Wiley & Sons, 2004.

[ONA 06] ONATSKI A., Determining the number of factors from empirical distribution of eigenvalues, Working paper, Columbia University, 2006.

[OSB 00] OSBORNE M.R., PRESNELL B., TURLACH B.A., "On the LASSO and its dual", *Journal of Computational and Graphical Statistics*, vol. 9, no. 2, pp. 319–337, June 2000.

[PAN 11] PANAHI A., VIBERG M., "Maximum a posteriori based regularization parameter selection", *Proceedings of the IEEE International Conference on Acoustics, Speech, and Signal Processing* (*ICASSP 2011*), IEEE, pp. 2452–2455, May 2011.

[PAS 05] PASCAL F., FORSTER P., OVARLEZ J.P., *et al.*, "Theoretical analysis of an improved covariance matrix estimator in Non-Gaussian noise: existence and algorithm analysis", *30th International Conference on Acoustics, Speech and Signal Processing (ICASSP' 05)*, vol. IV, Philadelphia, PA, pp. 69–72, March 2005.

[PAS 06] PASCAL F., Détection et estimation en environnement non-Gaussien, PhD Thesis, University of Nanterre/ONERA, December 2006.

[PAS 08] PASCAL F., CHITOUR Y., OVARLEZ J.P., *et al.*, "Covariance structure maximum-likelihood estimates in compound Gaussian noise: existence and algorithm analysis", *IEEE Transactions of Signal Processing*, vol. 56, pp. 34–48, January 2008.

[PAS 11] PASTUR L., SHCHERBINA M., *Eigenvalue Distribution of Large Random Matrices*, AMS, 2011.

[PAT 13] PATTON A.J., RAMADORAI T., "On the high-frequency dynamics of hedge fund risk exposures", *Journal of Finance*, vol. 68, no. 2, April 2013.

[POT 05] POTTERS M., BOUCHAUD J.P., "Financial applications of random matrix theory: old laces and new pieces", *Acta Physica Polonica B*, vol. 36, no. 9, pp. 2767–2784, 2005.

[RAC 10] RACICOT F.E., "Modeling hedge funds returns using the Kalman filter: an errors-in-variables perspective", *Atlantic Economic Journal*, vol. 38, no. 3, pp. 377–378, September 2010.

[RES] Hedge Funds Research (HFR), "HFR databases". Available at http://www.hedgefundresearch.com/.

[RIS 78] RISSANEN J., "Modeling by shortest data description", *Automatica*, vol. 14, pp. 465–471, 1978.

[ROL 77] ROLL R., "A critique of the asset pricing theory's tests part I: on past and potential testability of the theory", *Journal of Financial Economics*, vol. 4, no. 2, pp. 129–176, 1977.

[ROL 80] ROLL R., ROSS S.A., "An empirical investigation of the arbitrage pricing theory", *Journal of Finance*, vol. 35, no. 5, pp. 1073–1103, 1980.

[RON 08a] RONCALLI T., TEILETCHE J., "An alternative approach to alternative Beta", *Journal of Financial Transformation*, vol. 24, pp. 43–52, 2008.

[RON 08b] RONCALLI T., WEISANG G., "Tracking problems, hedge fund replication and alternative beta", 2008. Available at http://www. guillaume-weisang.com/Assets/Download/Roncalli _Weisang_HFR-PF-20081224.pdf.

[ROS 76] ROSS S.A., "The arbitrage theory of capital asset pricing", *Journal of Finance*, vol. 13, no. 3, pp. 341–360, 1976.

[RUC 10] RUCKDESCHEL P., "Optimally robust Kalman filtering", *Berichte des Fraunhofer ITWM*, number 185, Fraunhofer, ITWM, May 2010.

[SAD 10] SADKA R., "Liquidity risk and the cross-section of hedge-fund returns", *Journal of Financial Economics*, vol. 98, pp. 54–71, October 2010.

[SCH 05] SCHMIDT M., Least squares optimization with L1-norm regularization, Project Report number CS542B, UBC, University of Alberta, Canada, December 2005.

[SHA 64] SHARPE W.F., "Capital asset prices: a theory of market equilibrium under conditions of risk", *Journal of Finance*, vol. 19, no. 3, pp. 425–442, 1964.

[SHA 92] SHARPE W.F., "Asset allocation: management style and performance measurement", *Journal of Portfolio Management*, vol. 18, no. 2, pp. 7–19, 1992.

[STO 89] STOICA P., NEHORAI A., "MUSIC, maximum likelihood, and Cramer-Rao bound", *IEEE Transactions on Acoustics, Speech, and Signal Processing*, vol. 37, no. 5, pp. 720–741, 1989.

[STO 97] STOICA P., MOSES R., *Introduction to Spectral Analysis*, Prentice Hall, 1997.

[TIB 96] TIBSHIRANI R., "Regression shrinkage and selection via the LASSO", *Journal of the Royal Statistical Society, Series B*, vol. 58, pp. 267–288, 1996.

[TRE 62] TREYNOR J.L., "Toward a theory of market value of risky assets", Unpublished manuscript. A final version was published in 1999, in KORAJCZYK R.A., *Asset Pricing and Portfolio Performance: Models, Strategy and Performance Metrics*,Risk Books, (ed.) London, pp. 15–22, 1962.

[TRE 02] TREES H.L.V., *Optimum Array Processing, Part IV of Detection, Estimation and Modulation Theory*, John Wiley & Sons, 2002.

[TUK 60] TUKEY J.W., "A survey of sampling from contaminated distributions", *Contributions to Probability ans Statistics, Olkin and Others, Stanford University Press*, pp. 448–485, 1960.

[TUK 62] TUKEY J.W., "The future of data analysis", *The Annals of Mathematical Statistics*, vol. 33, no. 1, pp. 1–67, 1962.

[TYL 83] TYLER D.E., "Robustness and efficiency properties of scatter matrices", *Biometrika*, vol. 70, pp. 411–420, 1983.

[TYL 87a] TYLER D.E., "A distribution-free M-estimator of multivariate scatter", *The Annals of Statistics*, vol. 15, pp. 234–251, 1987.

[TYL 87b] TYLER D.E., "Statistical analysis for the angular central Gaussian distribution on the sphere", *Biometrika*, vol. 74, pp. 579–589, 1987.

Index

F

G

H

I